# Health and Safety in a Cha

When health and safety regulatory frameworks took their present form in the 1970s, they were seen as a triumph of welfare state intervention. Since then, as heavy industry has declined and office and retail employment have expanded, new ways of working have radically altered the context of health and safety policy. Many people have come to see health and safety interventions as an obstacle to innovation.

This book aims to address the changing context of health and safety policy, exploring concerns arising within the profession and the appropriate responses. Its manifesto for reform promises to frame the debate within the professional and policy community for a generation.

The result of a major research programme funded by the Institution of Occupational Safety and Health (IOSH), *Health and Safety in a Changing World* shows how health and safety policy has developed over time, how it is applied in practice and how best to make it fit-for-purpose in the 21st century. The book will be essential reading for professionals, practitioners and academic readers with an interest in the rapidly-evolving field of health and safety.

**Robert Dingwall** was Director of the Research Programme at the Institution of Occupational Safety and Health (IOSH).

**Shelley Frost** is the Executive Director for Policy at the Institution of Occupational Safety and Health (IOSH).

# Health and Safety in a Changing World

Edited by
**Robert Dingwall and Shelley Frost**

LONDON AND NEW YORK

First published 2017
by Routledge
2 Park Square, Milton Park, Abingdon, Oxon OX14 4RN

and by Routledge
711 Third Avenue, New York, NY 10017

*Routledge is an imprint of the Taylor & Francis Group, an informa business*

*British Library Cataloguing in Publication Data*
A catalogue record for this book is available from the British Library

*Library of Congress Cataloging in Publication Data*
Names: Dingwall, Robert, editor. | Frost, Shelley, 1972- editor.
Title: Health and safety in a changing world / edited by Robert Dingwall and Shelley Frost.Description: Abingdon, Oxon ; New York, NY : Routledge, 2017. | Includes bibliographical references and index.
Identifiers: LCCN 2016024159| ISBN 9781138944220 (hardback) | ISBN 9781138225213 (pbk.) | ISBN 9781315672021 (ebook)Subjects: LCSH: Industrial hygiene–Great Britain. | Industrial safety–Great Britain.
Classification: LCC HD7695 .H43 2017 | DDC 363.100941–dc23LC record available at https://lccn.loc.gov/2016024159

ISBN: 978-1-138-94422-0 (hbk)
ISBN: 978-1-138-22521-3 (pbk)
ISBN: 978-1-315-67202-1 (ebk)

Typeset in Bembo
by Servis Filmsetting Ltd, Stockport, Cheshire

# Contents

# Figures

# Tables

# Contributors

**Paul Almond** is Professor of Law at the University of Reading

**Nicholas Andreou** was formerly a Research Associate at the Centre for Organizational Health & Development, University of Nottingham

**Phil Bust** is Research Associate at Loughborough University

**Alistair Cheyne** is Professor of Organisational Psychology at Loughborough University

**Hilary Cowie** is Director of Research Operations and Head of Statistics Section at the Institute of Occupational Medicine

**Joanne O. Crawford** leads the Ergonomics and Human Factors section at the Institute of Occupational Medicine

**Andy Dainty** is Professor of Construction Sociology at Loughborough University

**Alice Davis** is a Research Scientist (Psychology) in Ergonomics and Human Factors at the Institute of Occupational Medicine

**David Denyer** is Professor of Leadership and Organisational Change at Cranfield University School of Management

**Robert Dingwall** is Director of Dingwall Enterprises Ltd and Professor of Sociology at Nottingham Trent University

**Noeleen Doherty** is Principal Research Fellow at Cranfield University School of Management

**Mike Esbester** is Senior Lecturer in History in the School of Social, Historical, and Literary Studies at the University of Portsmouth

**Aoife Finneran** was formerly a Research Associate at Loughborough University and is now Human Factors Specialist at the Rail Safety and Standards Board

**Mike Fray** is Senior Lecturer in Ergonomics and Human Factors at Loughborough University

**Shelley Frost** is Executive Director – Policy at the Institution of Occupational Safety and Health

**Alistair Gibb** is ECI Royal Academy of Engineering Professor of Complex Project Management at Loughborough University

**Jane Glover** was formerly a Research Associate at Loughborough University and is now a Research Fellow in the Business School at the University of Birmingham

**Elaine Yolande Gosling** was formerly a Research Associate at Loughborough University and is now an Ergonomics and Human Factors Consultant at PA Consulting Group

**Ruth Hartley** is Lecturer in Human Resource Management and Organisational Behaviour at Loughborough University

**Roger Haslam** is Professor of Ergonomics at Loughborough University

**David Hollis** was formerly a Research Associate at the Centre for Organizational Health & Development, University of Nottingham; and is now a doctoral candidate at the Open University

**Aditya Jain** is Assistant Professor in Human Resource Management at Nottingham University Business School

**Wendy Jones** is Research Associate at Loughborough University

**Stavroula Leka** is Professor of Work, Health & Policy and Director of the Centre for Organizational Health & Development at the University of Nottingham

**Jennie Morgan** was formerly a Research Associate at Loughborough University, and is now a Research Fellow in the Department of Sociology at the University of York

**Colin Pilbeam** is Principal Research Fellow at Cranfield University School of Management

**James Pinder** is Research Associate at Loughborough University

**Sarah Pink** is RMIT Distinguished Professor, and Director of the Digital Ethnography Research Centre at RMIT University, Melbourne

**Peter J. Ritchie** is Head of Information Science at the Institute of Occupational Medicine

**Guy H. Walker** is Associate Professor in Human Factors at Heriot-Watt University

**Patrick Waterson** is Reader in Human Factors and Complex Systems at Loughborough University

**Gerard Zwetsloot** is Honorary Professor of Occupational Health & Safety Management at the University of Nottingham

# Foreword

In recent decades, the world of work has changed significantly. Trade and industry have become globalized through improved communications, travel and transport, the reduction of tariff barriers and development within individual economies. Organizations are operating within more complex environments, where deregulation in some areas has been paralleled by the growth of new standards of practice and reporting in others. They are expected to take greater responsibility for managing and controlling their own risks. These changes have challenged traditional ways of thinking and working in the field of occupational safety and health. While there have been significant improvements, the stark fact remains that, globally, more than two million people die every year as a result of health and safety failures at their workplace. Health and safety professionals are pivotal in ensuring that relevant data and intelligence are translated by organizations into practical and proportionate actions.

The Institution of Occupational Safety and Health (IOSH) is committed to the development of the professional knowledge and skills to create workplaces that are safe, healthy and sustainable. We are the first European safety body to be awarded NGO status by the International Labour Organization (ILO). We have the status of a charity under English law, which means that our work must be directed towards the benefit of the public. Our objective is a world where work is safe and healthy for every working person, every day. Our members are bound by a code of conduct that meets public expectations of professionalism – integrity, competence, respect and service. IOSH promotes the recognition of occupational health and safety as key dimensions of responsible business practice, stimulating product or service quality, innovation and growth. In support of this goal, we systematically assemble relevant scientific and social scientific evidence for the benefit of all stakeholders in occupational safety and health. Where necessary, we fund and commission research to expand the evidence base for professional practice. We share this knowledge beyond our own members - to the whole community engaged with this work.

In 2008, IOSH decided to make its largest ever investment in research, through a programme of studies 'Health and Safety in a Changing World'. This was carried out, between 2010 and 2015, by teams from the Institute of Occupational Medicine, Loughborough University, Cranfield University and the Universities of Nottingham, Reading and Portsmouth. Detailed scientific reports on each project and on the programme as a whole are available through the IOSH website. This book provides a more accessible and less technical overview of the programme's work, and some of the actions that it has inspired. It will inform a wider public and professional debate about present approaches to employee protection, possible improvements, and the profession's contribution.

We are using this research ourselves to think strategically about how best to advance high professional standards, to engage with other stakeholders, and to enhance public trust in health and safety and those responsible for its delivery. The current development of a professional competency framework, available to both individuals and organizations, builds on the findings of this programme. It looks forward to a profession that can work more flexibly, creatively and proactively with others to identify and solve problems. Our members will be equipped with the skills to join the teams that are developing new workplaces and working practices as advocates for health and safety from the start.

We would personally like to thank all of the contributors to this publication – it provides an invaluable insight into the past, present and future OSH agenda, and its work will be transformational for our profession.

*Karen McDonnell*
*President, Institution of Occupational Safety and Health*
*November 2015–November 2016*


*Bill Gunnyeon*
*Chair, Board of Trustees*
*Institution of Occupational Safety and Health*

# Acknowledgements

This book represents the collective effort of many individuals over a period of almost ten years.

The original vision was developed by the IOSH Research Committee, which constituted a Programme Advisory Group – Tim Carter, Brian Kazer, John Burdett and David Walters – and sustained their support under successive chairs – Vince McNeilly, Graham Frobisher, Graeme Collinson and Vanessa Mayatt. IOSH support has been delivered through the various inputs of Luise Vassie, Jane White, Mary Ogungbeje, Andrea Alexander, Anne Wells and Richard Jones. Robert Dingwall has been assisted by Patricia Hulme, who consolidated and compiled the bibliography for this book. Numerous anonymous peer reviewers have also contributed, both to the commissioning of the original projects and to the evaluation of the final reports.

The research teams have been assisted by many individuals, groups and organisations who have generously supported their work and welcomed them into their workplaces during the course of the fieldwork. As is usual in such studies, they have all been assured of anonymity and confidentiality in published accounts of the work, except where they have agreed to be named, so we can only thank them in general terms. However, the projects have benefitted from the input of identifiable individuals through their various advisory groups and committees and we would like to recognize these people: Emma Doust, Richard Graveling, Judith Lamb, Marlyn Davis, Robert Atkinson, Steve Bailey, Sue King, Paul Litchfield, Sheila Pantry, Helen Pearson, Marion Richards, Hugh Robertson, Andrew Kennedy, Peter Fisher, Paul Haxell, Chris Jerman, Davide Nicolini, James Stapleton, Neil Stephens, Dylan Tutt, Lawrence Waterman, Carl Foulkes-Williams, Hayley Saunders, Glenn Sibbick, David Wallington, Graham Frobisher, Arthur McIvor, Sarah Page, Kevin Tesh, Neal Stone, Dianah Worman, Martin Bevan, Richard Booth, Trevor Dodd, Steve Doe, Andrew Foster, Chris Rowe, Joscelyne Shaw, David Smith and Ian Wrightson.

# Abbreviations

| | |
|---|---|
| ACoP | Approved code of practice |
| ALARP | 'as low as is reasonably practicable' |
| BOHS | British Occupational Hygiene Society |
| CDM | Construction (design and management) regulations |
| CORE | Cumbrians opposed to a radioactive environment |
| COSHH | Control of substances hazardous to health regulations |
| CSR | Corporate social responsibility |
| EU | European Union |
| HSC | Health and Safety Commission |
| HSE | Health and Safety Executive |
| HSWA | Health and Safety at Work etc. Act 1974 |
| IOSH | Institution of Occupational Safety and Health |
| KT | Knowledge transfer |
| LFS | Labour force survey |
| MHSW | Management of Health and Safety at Work Regulations |
| NEBOSH | National Examination Board in Occupational Safety and Health |
| NHS | National Health Service |
| OHS | Occupational health and safety |
| OSHCR | Occupational Safety and Health Consultants Register |
| OSH | Occupational safety and health |
| PPE | Personal protective equipment |
| RoSPA | Royal Society for the Prevention of Accidents |
| SHP | *Safety and Health Practitioner* [www.shponline.co.uk/] |
| SME | Small or medium enterprise |
| SRSC | Safety representatives and safety committees regulations |
| TUC | Trades Union Congress |

# Introduction

*Robert Dingwall and Shelley Frost*

This is a book about two visions. The first is the vision that every worker should finish their employment alive, in good health, and with the same number of limbs, digits or organs that they started with. This is a goal for every shift and for a working life as a whole. Who could possibly disagree with this? The means by which it can be achieved, however, have been increasingly questioned, in the UK and elsewhere, over the last 20 years. Worker health and safety have come to be seen as nice to have but, perhaps, something of a luxury in hard economic times. Looking after a workforce is not an investment in human capital but an overhead cost that reduces margins and business competitiveness. Although it may be difficult to argue against the principle that employment should not be a source of death, disease or disability, the institutions that give effect to this can be challenged. Rules and regulations can be characterized as too prescriptive and disproportionate. Enforcement can be depicted as petty and lacking in common sense. A framework devised for the industrial economy of the 1970s may be ill-adapted to the conditions of the service economy of the early twenty-first century.

The second vision arises in response to this challenge. The leading international institution representing health and safety professionals accepts the need to reimagine the profession's mission to keep in step with social and economic change. How can workers be kept safe and healthy when large-scale manufacturing organizations in heavy industries have given way to loosely connected networks of small and medium-sized enterprises engaged in light industry and service provision? What are the current hazards of employment? Is the regulatory system fit for purpose? What is the profession's role in this new environment? What skills will its members need to prepare them for work in years to come? The crisis of confidence in occupational safety and health is also a moment of opportunity, to define and create the professionals who will continue to serve as advocates for the lives and well-being of workers in the UK and around the world.

In order to inform that process, the *Institution of Occupational Safety and Health* (IOSH) decided, in 2008, to make its largest ever investment in research, through a programme of studies 'Health and Safety in a Changing World'. This Introduction describes the background to the programme and introduces the projects. It is followed by six chapters, contributed by the project teams, describing each in more depth. Finally, the Conclusion discusses what this work might mean for the profession and how IOSH itself has begun to respond.

## What questions did the programme ask?

The programme was designed to investigate the consequences of the changes that had taken place in the social and institutional context of occupational safety and health (OSH), with the restructuring of British industry and its workforce, the development of a degree of cultural hostility to perceived over-regulation and the attempt to promote an approach to workplace safety based on risk rather than on hazard. New approaches to health and safety practice were needed, reflecting the evolution of more networked forms of organization, the greater professionalism required to operate knowledge-based rather than rule-based assessments and interventions, and the need to secure public legitimacy for occupational health and safety as an endeavour. This led to the commissioning of projects around three specific themes:

### How did we get here from there?

The *Health and Safety at Work etc. Act* 1974 (HSWA) is often regarded within the profession as a high-water mark of its achievement. Although there has been growing interest among social scientists and historians in the processes that shaped this legislation, that story has not been brought together in a form readily accessible to the profession. If, however, the profession is to understand why its value has since been publicly questioned, it is essential both to know what made this legislation possible – and what has changed. The benefit of this work will be seen in recognizing how professional practice has not fully tracked changes in the nature of work and workplaces since the 1970s; in devising ways to realign practice and redesign its legal environment; and in identifying and enlisting potential allies to assert the continuing value of employee protection on the basis of a revised approach to intervention.

### What do we need to know?

Several of the criticisms currently being levelled at the health and safety profession centre on its relationship to knowledge. It is, for example, frequently alleged that health and safety professionals lack common sense. This is really a way of saying that they may have sound scientific or technical knowledge but are less good at making practical and proportionate use of this in specific cases. On other occasions, we may hear complaints about their poor judgement of risk, leading to unrealistic requirements about safety regimes. This is sometimes called 'gold-plating'. One small example might be the way recent changes in regulations about the testing of portable electrical devices are not actually being reflected in advice to small businesses. Annual testing of computers, heaters, kettles, and so on, may no longer be required but is still being recommended by consultants. A vision for the twenty-first century health and safety professional would need to include some specification of what that person needed to know, how they were going to keep their knowledge up to date, and how they were going to recognize and deal with their own limitations.

### How are we going to work?

The history of the health and safety profession is very much bound up with a world of large corporations, heavy industry and organized labour. Although there are still places where this model survives, contemporary economies are characterized by much looser forms of organization connecting relatively small enterprises in a joint effort to assemble some product or service. These workplaces present different sorts of hazard – a spillage of molten steel is replaced by repetitive injuries from assembling small components or stress from intensive performance management regimes. Union membership is increasingly confined to the shrinking public sector. Employment itself is more precarious and giving way to self-employment or microbusiness. The workforce is polarizing between groups of workers that may be more skilled and knowledgeable than an average health and safety professional on the specific hazards of their job and groups in low-skill work with little engagement beyond the earnings they can extract. In both sectors, many workers may be migrants, with a limited command of the local language and different cultural expectations about working conditions. If the overarching vision of a safe and healthy workplace is to be achieved, health and safety professionals must themselves find new ways of working. What does it mean to be a leader in an organization that has dispensed with many layers of management in favour of team-based structures? How can small businesses and individual workers be engaged with the vision of health and safety rather than simply being told to follow rules that may have little relevance to them? Where is it essential to stick to safety-critical protocols and where is it better to support local, improvised, solutions to problems that have been jointly identified?

## The social licence to operate

Policymakers traditionally assume that organizations approach issues like health and safety with a narrow focus on avoiding sanctions simply by complying with regulations. In practice, however, contemporary organizations often have to go beyond their minimal legal obligations and meet wider societal expectations about what is considered acceptable and unacceptable behaviour. They need a 'social licence to operate' (Gunningham et al. 2004). Organizations that do not pay enough attention to such concerns incur reputational and other costs that make it more difficult to function. This applies as much to regulators as to the firms that they supervise. A regulatory agency needs both economic, and social and cultural resources to use its enforcement powers. The outcomes are judged as successes or failures by a wider audience, in relation to their perceived proportionality and consistency with other values. Either failure or success may lead to a re-evaluation of the agency and changes in its legal powers or funding. Regulators and their target organizations operate within the same complex social and institutional system, which shapes the choices and actions that are available. This necessarily has a historical dimension: social scientists describe this as 'path dependence', the extent to which previous decisions constrain the options that are available at the present moment.

The first two chapters in this book explore different aspects of the history of the systems within which occupational health and safety is embedded. They investigate 'how we got here from there – and what does this mean for where we can go next'. In the opening chapter, Mike Esbester and Paul Almond focus on changes in the social, economic and cultural environment for health and safety interventions since the early 1960s and the way these have influenced perceptions of the legitimacy of these interventions. The second chapter, by Stavroula Leka, Aditya Jain and their colleagues, outlines the history of occupational health and safety regulation since its beginnings in the early nineteenth century, and identifies some of the specific drivers that have shaped this. They conclude by examining the engagement of contemporary stakeholders in the processes that are now shaping public policy and actions in relation to health and safety.

Occupational health and safety as we understand it today is very much the product of a particular historical period. Before the industrial revolution, workplaces were generally small-scale and technologies relatively simple. The main exceptions were a few enterprises closely associated with the military, like naval dockyards and cannon foundries. From the late eighteenth century onwards, an increasing proportion of British workers came to be employed in manufacturing and industrial production. Their workplaces grew in size, while technologies became more complex and introduced new health hazards. In the second half of the twentieth century, the structure of industry changed again: mass production moved to newly industrializing countries. The United Kingdom came to be dominated by service work and light industries, often serving specialized markets from relatively small production units. These sectors are sometimes thought to be low hazard, by comparison with traditional heavy industries like coal mining, steel making or shipbuilding, although this assumption is less straightforward than it seems.

Nevertheless, as an object of public policy, concern for occupational health and safety originated in an era dominated by large-scale industrial production, with low levels of automation. The workforce had limited education and economic resources, although it was capable of organizing for collective action. Employers varied greatly in their engagement with health and safety. Some were relatively supportive of regulation, either to avoid being undercut by less scrupulous rivals or because worker health and safety had implications for customer health and safety. Others regarded regulation as an unnecessary burden and an unjustified interference with freedom of contract. Although the extension of health and safety legislation, enforced by inspection, is often told as a simple story of historical progress, there has always been a degree of tension between competing views of the role of the state, the liberty of workers and the rights of owners or employers. The balance has shifted from time to time, particularly where the absence or weakness of regulation has been blamed for high-profile accidents or injuries, but there never was a Golden Age when health and safety was universally accepted in expansive terms as a national objective.

Many of the current issues in occupational health and safety policy and practice can be traced to the struggle to adapt institutions and practices designed for a society dominated by one mode of production to a society dominated by another. This process began in the 1960s, when the *Offices, Shops and Railway Premises Act* 1963 extended the nineteenth-century model of regulation to over one million additional workplaces, mainly in the service sector. With the obvious exception of the nationalized rail industry, many of these were small and medium-sized businesses. These enterprises had traditionally considered themselves to be low-hazard environments. If health and safety issues existed, they could be dealt with by straightforward self-protective actions taken by skilled and knowledgeable workers. In Chapter 5, Loughborough colleagues show that this is still true of many small businesses and microbusinesses.

This major expansion of intervention accentuated the problems that were emerging with the traditional approach, based on 'hard law' – specific prescriptions for behaviour inscribed in statute and backed by inspection and sanctions. Despite legislative consolidation in the *Factories Act* 1961, health and safety practice was based on a patchwork of regulations and inspectorates that had grown up in an *ad hoc* fashion, responding to particular problems in particular industries. This model had locked occupational health and safety into specific industry structures and production technologies. Limited inspection resources were consumed by routine and predictable schedules of workplace visits. The lack of resources meant that enforcement powers were rarely used. The system was widely criticized for failing to challenge either employers or workers about unsafe practices.

In 1970, the Labour government set up a committee under Lord Robens, a former trade unionist and Labour MP, who chaired the board of directors of the nationalized coal industry. This committee had a bipartisan and expert membership, reflecting the corporatist approach adopted by all leading politicians in the 1960s. This approach, which owed much to the successful example of German post-war economic reconstruction, envisaged that industrial policy and practice would be based on a tripartite collaboration between state, unions and employers. Some elements of this were contentious, such as worker engagement with decisions about company finances, investment and strategy. This was resisted by both employers and unions as a departure from their traditionally adversarial relationship. However, health and safety was thought to be an area where progress could be made, because it was assumed that employers and unions had a common interest in preventing injuries and illnesses.

The supposed unity of interest meant that detailed regulation could be replaced by more flexible 'soft law' instruments, such as codes of practice and voluntary standards developed through partnership working by stakeholders in particular industries. The 'hard law' approach of primary legislation was inevitably slow, ponderous and vulnerable to disruption by extraneous factors. Soft law could respond more quickly and flexibly to changes in industrial organization and production technologies. It might also be less controversial because it was created by the stakeholders themselves using what they considered to be authoritative expertise. A single, unified, inspectorate would facilitate these partnerships and adopt a more risk-based approach to enforcement. Inspection could concentrate on those enterprises that were slow to accept that a positive approach to health and safety was a defining characteristic of good management. The inspectorate would also act on behalf of the public to ensure that their protection was considered alongside that of employees.

The Robens approach received strong bipartisan support – the *Health and Safety at Work etc. Act* 1974 followed its recommendations very closely and was a minority Labour government measure, reintroducing a Conservative bill that had fallen with the dissolution of Parliament and a general election. However, its implementation ran into considerable resistance within the civil service and the inspectorates. The scale of organizational and cultural change required by the Robens model seems to have been underestimated. Opposition clearly persisted over a longer period than might have been expected. Moreover, implementation took place against a background of economic difficulties and a movement away from the corporatist thinking of the 1960s. The rapid decline of large-scale manufacturing industry was accompanied by a matching decline in trade union membership. This shifted the risk profile of British industry and the ability of both employers and workers to engage in the kind of soft law production that Robens had envisaged. The unified inspectorate acquired new duties in terms of public protection without corresponding mechanisms for public engagement. This left it out of step with wider demands for participation in shaping public services. Its generalist approach also seemed to be inadequate for dealing with some particularly high-hazard sectors: specialized inspectorates were subsequently reinvented for offshore oil and gas installations, the nuclear industry and the railways.

In practice, however, as both chapters note, rather limited use has been made of the flexibility offered by the Robens approach to regulation. Employers, especially at smaller scales, have continued to seek out prescriptions and checklists rather than use their expertise and judgement. As Judith Hackitt, former chair of the HSE, has observed in her blog (31 July 2015), the HSE's drive to simplify regulation and minimize its impact on small businesses has been matched by a growth in 'advice' that reintroduces paperwork. The reduction in 'red tape' has been accompanied by the spread of what others have called 'blue tape'. This refers to requirements imposed by insurers, banks and larger trading partners: even a one-person professional service company operating from a home office can be obliged to write a health and safety policy simply to satisfy the procurement rules of some public sector clients, although the law does not actually require a written policy from businesses with fewer than five employees. There is an expanding business in compliance, charging for services where self-help would be possible with the assistance of the HSE website, which has been extensively redesigned to facilitate this. Hackitt questions how far this leads to 'gold plating', over-elaborating risk management to justify the adviser's fee, and how far it is 'feather bedding', laying off risks that should be accepted by responsible and competent managers. She cites one example, but it is not hard to find others, where the complexity and uncertainty of regulatory requirements are exaggerated in order to sell a service (see also Sarat and Felstiner 1989 on the way lawyers present the law as a chaotic world in which clients need a trusty guide).

While the chapters do not directly address Hackitt's question, they suggest that 'red tape challenges' and similar initiatives are doomed to failure. Occupational health and safety risks need to be understood in a more complex social and institutional context, where a contraction in one area may simply lead to an expansion in another. Public, and publicly accountable, regulation may be replaced by private regulation that is every bit as burdensome but harder to debate.

Both chapters also underline the growing role of European legislation in the area of occupational health and safety. From a European perspective, a common approach to workplace health and safety is fundamental to creating a single market with uniform business overheads. Accordingly, UK ministers in the Conservative governments of the late 1980s and early 1990s seem to have traded extended European competence in health and safety against opt-outs in other policy areas. The European dimension has not been embraced with any enthusiasm by key actors – and is regarded as an example of over-reach and lack of accountability by some. Nevertheless, UK thinking on the merits of soft law approaches based on collaboration between stakeholders has had a considerable influence in remodelling the original EU tendency to rely on hard law and more prescriptive methods.

Where does this leave the social licence of the institutional system concerned with occupational health and safety? Esbester and Almond note the prominence of 'regulatory myths' in policy debates. When we look at public opinion surveys, however, we find a rather different picture. These surveys provide good evidence on public attitudes during the latter part of the period examined in Chapter 1. They show that attitudes to health and safety have been relatively stable. Respondents are consistently positive about the goals and outcomes of health and safety regulation, while being more negative about their sources. If people are asked to think about the deregulation of their own workplace, they are much less critical of health and safety interventions. Although Esbester and Almond's focus groups elicited various regulatory myths, the participants did not consider these to be typical: both focus group and survey data register a positive view of the health and safety inspectorate, for instance. The social licence to operate appears to be more resilient than we might think from the anecdotes in tabloid media or political speeches. Nevertheless, it is something that needs continually to be renewed as institutions, expectations and technologies evolve together.

## Knowledge and action in organizations

If the health and safety profession is to continue to be fit for purpose within changing forms of industrial organization and new production technologies, it will need to re-examine its own ways of working. Chapters 3 to 6 look forward to the mainstreaming of occupational health and safety practice within workplaces, rather than this being seen as a bolt-on. Health and safety professionals would move away from a role as shadow regulators, restricted to advising on compliance. The new vision sees them as problem-solvers, using their access to expertise in human and technical systems to achieve health and safety objectives within development and production teams. While health and safety professionals would remain guardians of the outcomes, in terms of minimizing both hazards and risks to workers, their practice would be better adapted to the speed and flexibility required for competitive innovation.

These four chapters examine the challenges that would be involved in shifting to this style of practice. Chapter 3, by Joanne Crawford and her colleagues, investigates the quality and accessibility of the explicit knowledge base that is used by occupational health and safety professionals. Chapters 4 and 5, by Loughborough colleagues, explore how occupational health and safety is accomplished within networked systems of production, with a particular emphasis on the issues facing small and medium-sized enterprises. Chapter 6, by Colin Pilbeam and his colleagues, considers the notion of safety leadership and how health and safety professionals can achieve positive outcomes, particularly in supposedly low-hazard environments.

### OSH knowledge and its translation into practice

Many regulatory myths centre on an, implicit or explicit, allegation that health and safety interventions are contrary to common sense. Because these interventions are not justified by robust evidence, they result in unreasonable, unrealistic and heavy-handed actions that are not proportionate to any possible benefits. One of the first questions for the research programme, then, was to ask how solid was the evidence base for professional practice in occupational health and safety. Could an evidence base be defined at all? Was it derived from good quality research? Just as importantly, was this evidence accessible? Could users, whether health and safety professionals or managers with health and safety responsibilities, build their practice on the most up-to-date information or was this locked away? Did they have to depend on outdated knowledge from earlier training courses, *ad hoc* experiences or professional gossip? How smoothly did knowledge move from research into practice?

As Chapter 3 shows, there is a large and diverse set of knowledge resources to support occupational health and safety practice. Most of this knowledge was, though, simply pushed out. The organizations that provided it knew relatively little about how their material was actually received or used. A 'professional' model of health and safety practice might imply that practitioners would be directly engaged with basic research. However, the range of expertise required by many health and safety professionals, unless they were employed in very specialized roles, meant that their capacity to benefit from basic research was quite limited. Although this knowledge was often behind paywalls in scientific journals, the main barrier to greater use was the time required to search for the information, evaluate it and translate it into relevant advice or actions. This problem was solved by knowledge brokers, intermediaries who scanned the primary research and adapted this into simpler packages for work with clients. These packages were evaluated in terms of their immediate practical application and the reputation of their source. It seems that it is not just enterprise managers who are looking for straightforward prescriptions, but also their health and safety advisers. This is, though, not a matter of laziness or indifference so much as an appropriate response to the range of topics that advisers might be expected to know about.

A fundamental problem for the profession is that the knowledge required to make an effective contribution to problem-solving may be too diverse to be managed by any single individual: a construction engineer is not necessarily also an expert on organic chemicals or human factors. Health and safety professionals face immediate and sometimes urgent challenges within a complex system of multiple and interacting variables. With career experience in a particular industry, a health and safety professional may be able to acquire sufficient knowledge to deal with many of these. Specialists working in larger companies also seem to maintain extensive personal networks that enable them to draw on other expertise as and when required. Some of these networks include former colleagues who are now working independently or in smaller companies and can tap into the networks when they need to. However, as the first two chapters show, the assumption that health and safety professionals will have stable careers in large and equally stable organizations is becoming increasingly less valid.

With a few exceptions, smaller companies do not have the resources to sustain specialist expertise or undertake extensive searches for knowledge, let alone assess its value when located. If they have an external health and safety consultant, that person may also be relatively isolated and have an intermittent or one-off relationship with the company rather than being a long-term partner who is regularly contributing to the development and strategic thinking within the business. If the industrial structure is increasingly marked by shifting networks of small and medium-sized organizations, the health and safety profession faces what social scientists would call a problem of distributed cognition, where successful performance of a task depends on pooling the knowledge and skills of multiple individuals rather than resting with a single person (Hutchins

1995). The solution is likely to involve the development of communities of practice that cut across conventional organizational boundaries and link health and safety professionals as members of a profession that shares knowledge (Lave and Wenger 1991). This will, though, need careful institutional management because of the obvious tension with more proprietary views of knowledge as an economic asset for an organization. In a sense, we come back to the Robens view that health and safety may be an area where there is more to be gained from collaboration, in this case between organizations, than from competition.

### Managing occupational health and safety in the new industrial structures

The traditional, high-hazard industries historically associated with occupational health and safety practice and regulation now play a much smaller part in the UK economy. The experience of these industries is well-documented and embedded in the experience of the health and safety professions. However, much less is known about how health and safety goals are achieved in the networked forms of organization that are more characteristic of contemporary industrial structures. Chapters 4 and 5, then, break new ground by their investigations, of larger-scale networks or network-type organizations, and of the, very small, microbusinesses that may be enrolled in them.

Chapter 4 examines networked organizations. These replace traditional forms of vertical integration and line management by project teams of workers collaborating on a service or production process but employed by independent companies. This may involve working across different sites and even countries, as in the supply chain for many supermarket products or fashion items, or workers may be brought together on a single site, as in construction. Their collaboration rests on a basis of contractual or quasi-contractual devices between companies of different sizes. It has been suggested that these relationships may provide an alternative to traditional forms of regulation and enforcement as the larger firms in the chain transfer knowledge and enforce quality standards to preserve their own social licence to operate. Upstream failures, including health and safety, may compromise the reputation of downstream partners: the impact of labour conditions in factories manufacturing clothing located in developing countries on consumer views of retailers in developed countries might be an example. The network approach has also infiltrated more traditional industries, particularly in retail and logistics, where it has flattened managerial hierarchies and increased the autonomy of local business units to respond to their specific market conditions (Timpson 2010; Tessier 2013).

Chapter 5 looks at small or medium enterprises (SMEs), which represent a growing sector of employment within the UK economy and which are thought to have particular difficulty in meeting occupational health and safety requirements. They feature prominently in regulatory myths and are regularly invoked by politicians critical of red tape. However, much of the debate fails to reflect the diversity of this sector: companies with 500 employees are very different from companies with five employees. At one end of this spectrum, we are dealing with small versions of big companies but at some point below that, there is a qualitative shift that may require a rather different approach to promoting positive health and safety outcomes.

These two chapters explore the way in which more flexible organizational models need to be accompanied by more flexible approaches to occupational health and safety. These are not necessarily well supported by conventional health and safety thinking. The rule-oriented approach of large, vertically-integrated firms could be replicated in some high-hazard contexts, notably construction and mining. Where there was a clear structure of lead and subcontractors, which is common in construction, the lead company could impose site rules on others. For some small companies, this was a source of trickle-down knowledge about emerging hazards and ways to deal with them. Others were just irritated and chose not to work on such sites.

This approach was, though, much harder to operate in contexts where line management involved the coordination of workers from different companies or with a degree of autonomy derived from their specific experience or expertise. In many contexts, management was diffused and workers were expected to take a good deal of personal responsibility for self-protection. This was not necessarily a matter of management indifference. In very small companies, the 'family' nature of the business culture might make owners reluctant to press long-term employees to work in different ways or to instruct them about how to deal with new hazards. In other contexts, management was at a distance: home delivery teams, for example, frequently had to work out safety issues for themselves as they confronted the configurations of particular properties and the demands or expectations of the residents.

These chapters draw attention to the high degree of successful improvisation that goes on in many workplaces. The people who are actually performing a task carry out their own assessments of risk and devise strategies for dealing with this. Traditionally, this has been seen as a problem for occupational health and safety regimes, which have often relied on employees following instructions without deviation. This model does not, however, work well in contexts where workers are better educated, more empowered and less directly supervised than would once have been the case. There are, of course, some particular high-hazard, high-risk environments where health and safety expectations do require close attention to prescriptive rules but there are relatively few of these – and even there it is desirable for workers to be able to identify when the rules are failing to deal with a specific challenge (Perrow 1981).

Both chapters point to the importance of tacit learning, of workers' sensitivity to their own physical and other limits and to the intrinsic benefits of studying, and learning from, 'work-arounds' rather than trying to ban them. These bottom-up approaches may be better supported by developing a role for occupational health and safety specialists as educators and consultants to the whole workforce. This may involve directly training workers in how to carry out risk assessments at their immediate worksite or helping line managers or work coordinators to provide such training. Both projects, however, also point to the problem of workers who are not connected to these opportunities to learn and share experience, either because they are employed on an agency basis or because of barriers of language and culture – these categories often overlap, of course. This is equally an issue for many micro or, as the term suggests, 'nano' organizations, employing five or fewer workers. These organizations can find themselves cut off from both formal and informal sources of health and safety knowledge, unless they happen to be members of a trade association with active interests in health and safety or can maintain network connections from previous employment. At this scale, there can also be a struggle to understand and absorb health and safety risk information, which often seemed to be designed to be read by specialists working within larger organizations. However, merely simplifying this risk information must be set against the extent to which checklist approaches generate the problems identified by Judith Hackitt.

These findings complement the discussions of knowledge flow, in Chapter 3, by identifying the importance of capturing the knowledge that already exists, or is developed, within the workplace, at whatever level. Part of the skill of health and safety professionals is to mediate between these different kinds of expertise and to combine them in developing working practices that are both intrinsically safe and likely to be implemented by those involved. They also need to support line managers in the skills of facilitating good quality *in situ* dynamic risk assessments.

Previous thinking about occupational health and safety may also have given insufficient recognition to the moral dimension of many health and safety roles. Chapter 5, in particular, underlines the 'family' nature of relationships between employers and employees, or business partners, at the nano scale. Like all families, this can be both a strength and weakness: a sense of moral obligation may lead workers to take inappropriate risks as much as lead owners or managers to make additional investments in protection. However, this also seems to be an element of larger networks. Key players want to avoid the reputational damage, and the implications for their social licence, that may arise from a record of apparent indifference to hazards and poor risk management. Being seen to act in a moral way may eventually become a driver for acting in a moral way. While there will always be rogue employers who will only be impacted by the risk of sanctions, it is arguable that the threat of shaming may be more important for many businesses, particularly at smaller scales. This would also be consistent with contemporary thinking in criminology on the potential of restitutive or restorative justice (Braithwaite 2002). Certain types of crime and offender may be better managed by a combination of public shaming and opportunities to make amends through apologies and compensatory actions acceptable to victims. While offenders are still morally censured, they are simultaneously offered a route to rehabilitation and reintegration with the community.

Many SMEs would like to do OSH well but are frustrated by their own lack of expertise, and the difficulty of obtaining this from consultants in a way that is helpful to them. As a result, both they, and their consultants, find themselves working within simplistic frameworks of rules, which, as Judith Hackitt noted, may impose substantial costs in compliance but seem ill-adapted to the business that is being regulated. The policy challenge is to build on the motivation to achieve a safe and healthy working environment while facilitating more creative ways to bring this about. This may involve focussing more on the desired outcome rather than on any specific route by which it may be achieved.

### Understanding leadership

Although the term 'safety leadership' has come into widespread use within the occupational health and safety community, this has not been closely linked with the growing body of work on leadership by researchers studying organizations. It might also be noted that the idea of *safety* leadership may implicitly exclude the possibility of leadership in relation to occupational *health*. Chapter 6, then, begins with a systematic review of the literature on safety leadership and compares this to the wider scholarship on leadership in general. Safety leadership proves to be an ill-defined concept. It originates as part of a top-down, command and control, model that has been increasingly discredited as a basis for effective organizations. The weaknesses of that model are reflected in the movement towards various forms of networked organization, which we have already discussed.

Although it had evolved somewhat subsequently, the concept of safety leadership retained an individualistic orientation. There is little consideration of what it takes to get people to follow the leader: leadership is an interaction within a system of social relationships, not a fixed attribute of a role. In one direction, this has led to a bland assertion that everyone is a safety leader – which creates the risk that no-one leads and everyone assumes that action is someone else's responsibility. In another, the identification of specific safety leaders or champions tends to turn safety into an activity for specialists rather than an integral part of the everyday life of the workplace, as we also see in Chapters 4 and 5. This does not mean that there is no place for specialists as advisers or consultants with expert knowledge: it does, though, question the assumption that safety as a goal can be managed separately from management towards other organizational goals.

Contemporary work on organizations favours more plural or distributed forms of leadership. These see leadership as a collective activity, shared by agreement within a team as particular problems or challenges arise. As the authors note, this is particularly relevant in the large number of relatively low-hazard environments presented by the economies of developed countries, where team working is common. Leadership is seen in a more dynamic way, responding flexibly and adaptively to hazards as they arise. Again, this echoes many of the findings reported in Chapters 4 and 5. Leadership may be more a matter of reflective practice in the moment than of generic rules and policies transmitted from head office without regard to local contingencies. On a cautionary note, this may leave a gap in relation to strategic thinking about the occupational health and safety challenges facing an industry and in responding to emerging issues, such as psychosocial hazards. Organizations still need to think about their connections to the knowledge flows discussed in Chapter 3.

Chapter 6, then, presents findings from a series of empirical investigations into the practice of leadership in four organizations that might be considered to be low-hazard. The findings underlined the extent to which safety management in these environments was a very local issue, where relatively junior staff had the primary responsibility for identifying and managing hazards. Although company management could create a context through their standards and guidance notes, this was important mainly as a signal about the value attached to safety outcomes rather than being directly applicable in practice.

Hazards were typically local and contingent, deriving from the vagaries of store design, location and clientele. There was little cumulation of knowledge about hazards, management and prevention. Problems were typically fixed on the spot and the workers moved on to the next task. Near-miss reporting significantly understated the incidence of hazards so that organizational learning was limited. As a result, similar incidents could recur and pass largely unnoticed. However, it was clear that employees approached safety in a positive fashion. They were committed to their own protection, to the protection of co-workers and to the protection of customers. If anyone identified a hazard, they were likely to take some leadership action in managing it. What they were less good at was following through with action intended to anticipate and prevent the recurrence of the hazard in their own or other sites.

The project also identified the risk that marginal workers would be less engaged in sharing knowledge about hazards, as previous chapters have also noted. Each workplace had an identifiable core of workers, supervisors and local managers who were densely linked but it was also possible to identify other workers who might be described as semi-detached. It may seem self-evident that a construction site with a large number of workers who do not have English as a first language, and are employed through agencies, may have a problem with enrolling them in an effective safety regime, as Chapter 4 notes. Such issues may be equally salient in an apparently homogenous workforce, where some members are less integrated than others for reasons that may not appear immediately obvious.

## Conclusion

In some respects, this research programme is the most comprehensive review of occupational health and safety since the Robens Report. Its conclusions underline just how much the world of work has changed in the last 40 years. They begin to explain many of the tensions that have emerged as a model of regulation and practice designed for one epoch of economic and industrial organization has tried to keep pace with the emergence of another. These strains have shown themselves in challenges to the legitimacy of health and safety interventions, to the expertise and competence of the professionals involved and to the profession's adaptation to the new organizational forms within which its skills are deployed. This book is not designed to deliver solutions to these problems: indeed, its spirit is very much that solutions can only be found by the parties themselves, working collaboratively to develop appropriate means of ensuring that every worker does indeed go home safely. However, the contributors have substantially clarified the problems, defined the parameters within which solutions are likely to be found and identified ways of thinking that are capable of leading to efficient and effective professional and organizational strategies. In itself, this programme exemplifies the shift from a time when occupational health and safety was seen as a matter for top-down impositions to a time when it has become a shared goal, whose accomplishment rests on the combined knowledge and creativity of people at all levels of an organization, supported by the expertise of safety professionals.

# 1 Do the public have a problem with health and safety?

*Mike Esbester and Paul Almond*

## Introduction

Most people would like to think that their job makes a useful contribution to society. Surely there are few things more worthwhile than ensuring that every worker goes home safely at the end of their shift, or that citizens are able to enjoy services, events or everyday activities without risking their lives unnecessarily? It comes as something of a shock, then, when members of the professions and organizations concerned with health and safety find themselves on the end of political and media criticism as lacking common sense and obstructing growth and competitiveness in UK industry. They may look back to a time when the *Health and Safety at Work etc. Act* 1974 (hereafter HSWA) was passed, apparently with the support of a broad public and policy consensus. Now they are the object of reviews, by a former industrialist and Conservative minister, Lord Young of Graffham (2010), and a Swedish academic working in London, Ragnar Löfstedt (2011a), intended to remove 'red tape' and reduce the 'burden' of regulation.

How did we get 'here' from 'there'? What has happened over the last 50 years to undermine the view of the HSWA as a high-water mark of social progress? How has *'elf 'n safety'* come to be treated with derision? What can the health and safety professions and organizations learn from that experience to help them develop appropriate policies in the future?

## Background

The challenge faced by the health and safety profession is one known to social scientists as a problem of *legitimacy*. We can think of this as the respect that is given to some institution, like the health and safety profession, by other groups that it deals with. In more formal language, is there a 'generalized perception or assumption' that organizations or professions concerned with health and safety are 'desirable, proper, or appropriate within some socially constructed system of norms, values, beliefs, and definitions' (Suchman 1995: 574)? Judgements about this appropriateness or acceptability are usually made with reference to two types of criteria: those that are procedural (about whether due process and accountability are present), and those that are moral (about whether the outcomes pursued are right and just). Social scientists sometimes talk about this system as a *social licence to operate* (Gunningham et al. 2004). The idea of a licence goes beyond the narrow view that an institution should only be concerned with the search for profit through efficiency, as indifference to the wider social impact of an institution's activities may actually undermine its economic goals. An institution that is not generally regarded as legitimate finds it harder to function. The lack of respect makes it more costly to secure resources, implement policies and persuade other institutions to cooperate in joint activities.

Our project, then, took 'health and safety' as a broad area of activity, ranging more widely than the activities of regulators like the *Health and Safety Executive* (HSE) or the formal legal frameworks (such as the HSWA) which govern the area. This meant examining developments such as the emergence of a safety profession and of new stakeholders around health and safety, the changing nature of the British workplace, general social and political trends and the emergence of new types of risk. We used a mixed-method approach to document the ways that health and safety has been understood since 1960. Three kinds of data were collected. First, in order to understand the policy judgements made at different times, we conducted 40 oral history interviews with key surviving stakeholders (regulators, safety professionals, policymakers, trade union officials and employer representatives) from the field of health and safety. These interviews were semi-structured: the interviewer prompted interviewees to talk about specific topics but encouraged them to shape the discussions according to their own recollections of their role in events. Second, we examined a wide range of historical, archival and documentary sources produced by state bodies, trades unions, employers' organizations, workers, the media and non-governmental organizations such as the *Institution of Occupational Safety and Health* (IOSH), the *British Safety Council* and the *Royal Society for the Prevention of Accidents* (RoSPA). These records were initially sampled to find out what was available: particular areas were then selected for more detailed investigation. Third, colleagues at the University of Nottingham (Jain and Leka 2015) searched for available survey data on public attitudes to health and safety, which was mostly from the latter part of the period. These were complemented by a series of focus groups (n = 8) which we conducted with a representative sample of members of the public (n = 67) to discuss contemporary attitudes towards health and safety.

A detailed account of the development of the institutional framework around health and safety regulation in the UK can be found in the next chapter. Here we focus on the social, economic and cultural contexts within which health and safety regulation has been created and implemented. How have they shaped the social licence of the professions and organizations involved? We outline eight critical moments in the story of health and safety since 1960. Although there is not the space here to go into great detail about each, they will introduce 14 key themes which characterize the changing perceptions of legitimacy over this period. Any attempt by the health and safety professions to influence the way that others think about them will need to take account of all of these themes.

First, though, it is worth underlining the findings from the survey evidence and our own focus groups: throughout this period, as far as we can establish, public attitudes towards health and safety regulation have generally been strongly positive. These are issues that matter to ordinary people, particularly when there are serious consequences or risks involved. A moral argument for safety regulation is widely recognized (see also Almond and Colover 2012; Walls et al. 2004). The focus groups established that negative opinions about specific aspects of health and safety are laid over basic positive appreciation of the need for intervention. There is a 'critical trust' in regulators and 'safety people' (Walls et al. 2004): tendencies towards rigidity, risk aversion and officiousness were unpopular, although being serious, strict and officious also increased trust. The most negative aspects of public perceptions (dislike of commercial motives and the 'compensation culture', lack of respect for expertise) relate both to wider themes identified below, and to an increasing sense of disengagement from, or lack of involvement in, 'health and safety' (Jain and Leka 2015).

## Eight critical moments in health and safety

### *'Such a thing is inevitable': the Flixborough explosion, 1974*

One of the most dramatic instances of workplace dangers since 1960 was the Nypro chemical plant explosion at Flixborough in Lincolnshire in 1974. This caused widespread damage: 28 people died and 36 more were seriously injured. It was a highly visible demonstration of the ways in which workplace hazards might affect the public. Flixborough followed a series of cases during the 1960s and early 1970s that had prompted recognition of the public risks associated with industrial and commercial activities. These included the collapse of a mobile crane at Brent Cross in London in 1964 (killing seven and injuring 32) and the collapse of a spoil heap at Aberfan in South Wales in 1966 (killing 144, mostly schoolchildren). The Factory Inspectorate (one of the regulators merged into the HSE in 1974) was increasingly concerned about the scale and development of new processes (particularly chemical hazards). In his 1967 annual report, for example, the Chief Inspector of Factories noted 'the increased scale of modern manufacture, the vastly larger plant used, the higher speeds of much machinery [and …] the storage and use of very large quantities' of potentially hazardous materials in close proximity to residents (Crown 1967: xiii).

The ignition of a massive, unplanned, hydrocarbon release at Flixborough gave concrete expression to these fears. Although those killed and injured were all employees, there was extensive property damage within a three-mile radius. The 'holocaust at Flixborough' received widespread media coverage (*Guardian* 1974: 10). The event occurred while the HSWA was passing through Parliament (the Act received royal assent nine weeks after the explosion). Janet Asherson, formerly of HSE, recalled:

It was very much a public issue because we had that ghastly explosion [...] that pushed forward the development of the Health and Safety Work Act in 1974. [...]. There was a huge sea change in legislation. It was filling a big gap.[1]

The public and media reaction added weight to what would become Section 3 of the HSWA, imposing a duty on the HSE to ensure the safety of those beyond the workplace. This has become an increasingly significant part of the HSE's role. Nevertheless, these means of managing the risks to workers and the public created by modern industrial processes had limitations. In a 1977 interview, the then plant safety manager for Flixborough concluded:

we as a whole must accept the view that chances are, that somewhere, sometime there will be a major industrial incident. [...] such a thing is inevitable in the sort of times we are living with our needs for a technological society.[2]

The Flixborough explosion exemplifies the changing nature of risks and the way that pre-1974 regulation was failing to keep pace. It also demonstrates the lack of public engagement in debate about the acceptability of technological risks. No one had asked the villagers living around Flixborough whether the explosion risk was acceptable to them, and efforts to engage the wider public with issues of risk regulation were generally limited at this time.

### 'No one knew it was dangerous':[3] asbestos and health, 1960–2015

In contrast to the sudden, massive and public nature of events like the Flixborough explosion, occupational health issues have rarely provoked significant public awareness. As the Chief Inspector of Factories observed in 1969, though:

it is now clear that in many areas [...] for some groups of worker the risk of injury or death from industrial disease is much greater than that from accidents [...] at least as much effort must be put into the control of toxic contaminants in the atmosphere as into the elimination of the physical causes of accidents.

(Crown *c*.1969: xiii–xiv)

The differential treatment of safety and health remained, however. There are many reasons for this disparity, including the scientific and legal difficulties in establishing causation, the individualized nature of ill health and the long latency period before many diseases manifest themselves. Occupational ill health rarely produces 'newsworthy' events, as the former Chief Inspector of Factories, David Eves, noted:

something like the train accident at Ladbroke Grove, national front page headlines, public enquiries, everybody understands it was a terrible thing [...]. Something like asbestos, it's a much slower kind of burn. We've known about problems with asbestos for 100 years, but the regulation of asbestos has been spasmodic.[4]

Rather than a single 'moment' which shone a spotlight on the issue, asbestos only garnered occasional public and political attention – notably around the Acre Mill factory in Hebden Bridge in the 1970s and early 1980s, including the TV documentary *'Alice – a fight for life'* (1982). The state-appointed *Advisory Committee on Asbestos* (1976–79) attempted to 'rais[e] the standard of public debate and the level of public understanding' (Simpson 1977: 1). For health issues, like asbestos-related disease, there remain confounding factors that limit public debate. The science behind asbestosis, for instance, emerged slowly and was contested (particularly by industry lobby groups) through highly technical arguments. Both of these factors tend to be off-putting to the public.

The extent to which workers were, and are, aware of the health risks to which they are exposed is also highlighted by the case of asbestos. Testimony from people employed from the 1960s into the 1980s suggests that 'no one knew it was dangerous'.[5] Consequently, employees were unable either to protect themselves or to fight to reduce the danger. Nevertheless, there is also strong evidence that 'danger money' induced workers to trade health for wages (even if they were not fully aware of this bargain).[6] Although it is now less common to buy off workers in this way, other sorts of negotiation around health and safety issues still occur. Moreover, asbestos-related ill health does not simply affect those who have worked with it. Families were exposed to asbestos fibres on work clothing; asbestos was used in a huge variety of places; and dust escaped from factories such as Acre Mill or was released in the process of asbestos removal. Dan Shears of the GMB union criticized a recent HSE campaign aimed at getting small building firms to work safely with asbestos on the basis that 'what you're seeing is a lot of people who do work for larger businesses, "don't worry so much about this asbestos, just make sure you've got the right gloves and PPE" [...]. That's quite contradictory to what we would say',[7] namely, that there was no such thing as safe work with asbestos.

The asbestos case illustrates the uncertainties and difficulties in dealing with health issues, which may take a long time to appear, and in identifying risks that cross 'the factory fence'. While it also identifies the important role played by trade unions, it shows the problems of dealing with health and safety in non-unionized or weakly unionized sectors as well.

### *'A hazard continually kept in the public eye': nuclear power and the public, 1956–2015*

Civilian nuclear power has been one area of visible public debate about health and safety risks. As with asbestos, these risks diffuse well beyond the workforce. However, they have been viewed as particularly insidious threats to both safety and health because of their intangibility. Nuclear power genera-tion started in the UK in 1956. It continues to the present day, with nine opera-tional sites and ten retired. Although health and safety concerns have been evident throughout the UK's nuclear programme, they have become particularly pro-nounced since the 1970s. Both national groups, such as *Friends of the Earth* and *Greenpeace*, and local residents have increasingly challenged the nuclear indus-try on the robustness of its health and safety protocols. The legitimacy of regu *Inspectorate* (which came under the HSE) to questions in 1976, from the Secretary of State for Energy and others, was criticized as 'evasive' – whilst also revealing potentially dangerous incidents – not a confidence-inspiring combination (Wright 1977: 2).

The planning inquiry for the Sizewell B power station (1982–87) demon-strated that both regulators and industry had to take account of public concerns. As with asbestos, the scientific detail was difficult to understand and did not address questions that troubled the public. As a result, it was proposed during the inquiry that HSE should explicitly assess 'tolerable levels of individual and societal risk to workers and the public' in relation to nuclear power (Layfield 1987: summary). The HSE went on to develop the 'Tolerability of risk' schema, which included a place for judgements about the social acceptability of risks, alongside traditional scientific criteria. As the then chair of the *Health and Safety Commission* (HSC) observed, during select committee hearings in 1988: 'technological change is now probably swifter [...] the public has become increasingly conscious of and knowledgeable about its implications. The reas-surance provided by a fully effective and respected state regulatory body is more and more important' (Crown 1988: 2). The tolerability of risk frame-work became a highly influential document and was revised in 1992 and 2001 to make it applicable to a broader range of industries (Bandle 2007).

It is important to acknowledge the diversity of opinion in relation to living with nuclear risks. Opposition groups such as *Cumbrians Opposed to a Radioactive Environment* (CORE, established 1980) were sometimes able to pressure the HSE into acting, as with their 1988 dossier about alleged malpractices at Sellafield.[8] A contributor to *The Times* commented, in 2003, that the HSE 'have in part been responding to a public perception of a hazard that, although small, has continually been kept in the public eye by pressure groups' (Spare 2003: 23). Yet many living in the immediate locality of nuclear power have to develop understandings of the risks involved that enable them to remain, as was noted of the population around Dounreay in Scotland:

the vast majority of the local people [...] were much more aware of the potential risks and the fact that the risks had been maintained at a sensibly low level over the years. So it's extremely helpful having a local popula-tion [...] who are able to give reassurance to those who otherwise might feel much more concerned about it.[9]

In contrast to asbestos, the nuclear case shows that even long-term and diffuse risks, to both workers and the wider community, can become the subject of public concern and pressure. Alternative sources of expertise developed, challenging the competence of industrial scientists and safety professionals. 'Disinterested' NGO expertise was both accessible and useful to local populations, in contrast to the expertise of those with a stake in the nuclear industry, which struggled to communicate a clear and obviously disinterested message to these constituencies.

### 'The wonderful day when we have Worker Safety Representatives': establishing formal employee representation after 1970

Since the First World War, both the state and employer organizations had called for the voluntary introduction of safety committees – with limited effect: a Factory Inspectorate survey in 1967 discovered that only 33 per cent of SMEs (50–500 employees) had joint safety committees (Crown *c*.1969: 1). Where these were established, they did not always include worker representatives. Unsurprisingly, unions were dissatisfied with such informal arrangements: the TUC noted in 1968 that 'existing safety organisation reflects the piece-meal nature of its development and is not capable of meeting present day needs',[10] and urged compulsory worker representation in evidence to the Robens *Committee on Safety and Health at Work* (1970–72). But Robens (1972) did not accept this, recommending instead a duty to consult employees. This was followed in the HSWA, to the disappointment of the unions, who then pressed the idea of representation on a Labour government expected to be more sympathetic to the principle. Despite concern from employers about worker involvement in 'managerial' matters, the *Safety Committee and Safety Representatives Regulations* were passed in 1977. As David Eves, later Deputy Director-General of the HSE, recalled

> you want [...] employers, and employees' representatives, usually through the trade unions, to agree that there is a way forward that both sides can agree to. So this is why the Safety Committee and Safety Representatives Regulations were one of the Commission's first priorities in the 1970s, and they struggled with it at first, it took them three years [...]. These for the first time gave safety representatives who were elected by their fellow workers, powers to inspect workplaces themselves, investigate, bring matters to the attention of their bosses.[11]

Some remembered an initial flush of enthusiasm, muted by the fact that the Regulations depended upon trade union representatives, which created problems where unions were weak: '76 safety reps were being trained and there was a huge enthusiasm. I think in the unionized companies that structure worked extremely well, in the non-unionized companies sometimes there was very little interest'.[12] The limitations were further exposed as trade union membership and traditional industries declined during the 1980s: newer sectors, less likely to be unionized, became a bigger part of the economy. It could be difficult to attract people to fill what might be demanding roles. Workers also expressed concerns about the sincerity of employers in instituting representation, as in the bitter 1979 comment by a Yorkshire farmer that

> [l]ong before the wonderful day when we have Worker Safety Representatives, [the HSWA] must be gone over by a fine toothcomb and brought up to the standard of the Criminal and Common Law, otherwise "barrack room lawyer" employers will laugh many of these Worker Safety Representatives off their farms.
>
> (Tubby 1979: 6)

Similarly, in 1980, the shop steward for one manufacturing firm was 'disappointed with the Company's response to the safety representatives' reports, it seems that requisitions go in but they have no intention of doing many of the jobs unless we draw special attention to a specific hazard'.[13] This case also recognizes the important role played by trade unions and the problems of dealing with health and safety in non-unionized or weakly unionized sectors. While asbestos related to a specific hazard, this brings out the more general issues presented by the shifting balance between worker organization and management authority.

### *'I still get emotional about it now … everything was shocking and it lasted for a very, very long time'*[14]: Piper Alpha, 1988

During the night of 6 July 1988, on the Piper Alpha oil production platform in the North Sea, a temporary valve fitted to a faulty pump failed. This released gas which triggered a colossal explosion, ripping the platform apart and leading to the deaths of 165 (of 226) crew members, and two rescuers. It remains Britain's worst industrial disaster of the last 80 years. Piper Alpha had a number of effects upon health and safety in the UK, both at the time and subsequently. First, it reopened the debate on safety standards and practices in the North Sea at the end of the 1980s. The disaster was caused by a multitude of failures in safety provision and management within Occidental, the company operating the platform, not least a lack of proper planning around safety management and maintenance work, and a lack of training and safety protection (including working fire extinguishers). This exposed wider problems within the sector and increased concerns about safety; Peter Jacques, then of the TUC, recalled: 'Piper Alpha actually kicked health and safety back into the centre of […] industrial life, and maybe […] political life'.[15]

Crucially, however, it also underlined a trend identified by Carson (1982): the subjugation of safety considerations to the overwhelming pressure for profitability and production within the offshore industry. One reason for the devastating nature of the Piper Alpha explosion was that gas supplies from other North Sea platforms were not switched off because of the prohibitive shutdown costs: the flames continued to be fed after the first explosion because production could not be compromised until it was clear a catastrophic breakdown had occurred. Questions about profit prioritization also emerged around the regulatory system that oversaw offshore safety. The Department of Energy was responsible for this, while simultaneously being expected to maximize sector revenues: this conflict of interest was indicted by the subsequent investigation (Cullen 1990) as contributing to a lax approach to regulatory oversight. The HSE was subsequently given responsibility for offshore safety, and regulation moved onto a more rigorous, safety case-based, footing. Piper Alpha (and other events, such as the *Herald of Free Enterprise* sinking in 1987) 'protected' the HSE, and health and safety regulation more generally, from government interference. They made deregulation politically untenable and reinforced the need for state protection against health and safety risks. Piper Alpha also had a significant impact on other industries. It led to the development of expertise in safety management systems and goal-setting approaches, which other sectors (such as the rail industry) would draw on, and benefit from, in dealing with their own disasters.[16]

Piper Alpha takes forward the issues raised by the nuclear industry case about the perceived tensions between health and safety, on the one hand, and company profitability on the other. It also makes clear that the political acceptability of health and safety regulation, as a whole, has unfortunately tended to be contingent upon the kind of expedient pressures and concern that attaches to the memory of recent and highly damaging safety failures.

### *'Bleeding ridiculous': office work, VDUs and health and safety after 1980*

Office work was, from the 1980s, an increasingly significant part of the UK's economy. Although its hazards were relatively limited, they were pervasive. Consequently, regulation started to touch large numbers of people who had not previously encountered it, with important repercussions for the legitimacy of health and safety. Health and safety in office work had (under the catch-all term 'welfare') received attention since at least the start of the twentieth century, usually around factors such as illumination and temperature. *The Offices, Shops and Railway Premises Act* 1963 extended formal regulatory oversight (by local authorities), while still suggesting that health and safety issues in offices were seen as rather less significant than the hazards posed by more traditional heavy industries. This legitimacy problem was to sharpen after 1992. Computer technology was not entirely new to office work in the 1980s. As early as 1971, it had been recognized in evidence to the Robens Committee, whose recommendations formed the basis of the HSWA, that: '[w]ith the development of computerized manufacture the border between the conditions of work in the factory and those in the office has become very much blurred'.[17] However, computer technologies, including visual display units (VDUs, or display screen equipment, DSEs) spread dramatically in the 1980s. They prompted public fears about unknown dangers that were being involuntarily assumed by workers. As well as musculoskeletal injuries, stress and fatigue,[18] one major area of debate concerned radiation exposure from VDUs and their potential impact on women's reproductive health (Denning 1985: 15), purported links which were later demonstrated to be unfounded.

Nevertheless, these concerns resulted in the introduction of *Health and Safety (Display Screen Equipment) Regulations* in 1992, as part of the 'six-pack' of European regulations. Given the increasing proportion of the workforce employed in office-based roles, these regulations affected a large number of people. Lawrence Waterman, a leading health and safety practitioner, observed that they were *'bleeding ridiculous to be honest'.*[19] They were felt by many, including HSE staff, to be disproportionate to the risks involved, and hence produced some discontent about health and safety in general. As one former HSE policymaker reflected, 'we've been pushed into regulating some comparatively low risk areas which have high impact on people. I've always had slight worries about the Display Screen Equipment Regulations, for example [...] it's a big step to say that every individual work station used anywhere in the country, must be properly assessed by a competent person in order to ensure the safety of the operator. And that impacts on more or less everyone at work and people think, "f*** this health and safety, it's a real pain, I just want to get on with my job."'.[20] This is not to deny that there have been, and remain, serious risks associated with VDUs and office work more generally.

The DSE case brings out the growing influence of EU regulations on UK health and safety practice, which will be explored further in the next chapter. It also underlines the country's changing industrial structure as employment shifted from traditional heavy industries to (apparently) low-risk office and service-sector work, bringing many people into contact with 'health and safety' for the first time.

## *'An absolute disaster': rail privatization, Ladbroke Grove and public debate, 1994–2015*

Much more immediately visible (and so more political) than office health and safety was railway safety in the post-1994 privatization era. Since the origins of the railway system, the public have been exposed to industrialized risks over which they have little or no control. Within a privatized system, as was the case in the nineteenth century, state oversight was deemed necessary: *Her Majesty's Railway Inspectorate* was formed in 1840. It endured for 150 years, remaining broadly outside the changes imposed by the HSWA and only being taken into HSE in 1990 following the Clapham Junction crash of 1988, although it effectively remained functionally separate. Within the HSE, it oversaw the privatization of British Rail (BR) between 1994 and 1997, when the nationalized system was broken up and sold off, with a number of franchises operating the services and another maintaining infrastructure. Privatization was predicated on the introduction of safety cases, following the nuclear and offshore licensing regimes, a massive culture change for the railway industry. According to Jenny Bacon, then Director-General of the HSE, privatization was 'an absolute disaster from a health and safety point of view [...] we ended up with 128 different parties, trying to manage interfaces and get them to talk to each other about health and safety issues'.[21] It introduced dynamics of contracting and subcontracting, stronger commercial imperatives to cut costs, and a loss of institutional memory as former BR employees left the industry, all of which had safety implications.

These manifested themselves in a series of high-profile crashes and derailments in the late 1990s and early 2000s, including Southall (1997), Hatfield (2000) and Potter's Bar (2002), all of which had an impact on public perceptions of railway safety and the HSE. However, the 1999 Ladbroke Grove crash, in which two trains collided killing 31 and injuring over 520, posed particular challenges to the HSE and HSC's legitimacy; one former HSE policy advisor noted that Ladbroke Grove resulted in public questions

> about the effectiveness of the Commission and more particularly the Executive in terms of carrying out its responsibilities. So, it wasn't just battle over red tape or resources. There was an even bigger political battle in terms of the structure and organisation of HSE and what it should have oversight of [...] one of the consequences was that railway inspectorate was moved out to what is now ORR [the Office of Rail and Road].[22]

This subsequent break-up of the HSE (including, in 2014, loss of nuclear regulation) has been perceived as extremely damaging to the HSE's legitimacy, challenging its status and competence and undermining the notion of a single, unified regulatory regime, which had been established by the HSWA.

These rail accidents were a strong reminder that traditional industries remained important sources of risk. Fairly or not, it was thought that the privatization of the railways had introduced commercial pressures to cut corners on safety that had undermined the risk-averse approach that had been embedded in the nationalized industry. At the same time, some of the consequences of these events, in terms of prolonged periods of service disruption, were derided as overreactions – slamming stable doors long after the horse had bolted.

### 'You hear "conkers in the playground" ... Is that true though?':[23] media, myth and politics, 2004–2015

In 2004, when head teacher Shaun Halfpenny introduced a health and safety rule that pupils at Cummersdale Primary School in Cumbria must wear safety goggles to play conkers, he claimed that he was 'just being sensible. We live in a litigious society'.[24] The story's novelty value meant it was quickly picked up by the national news media. It has subsequently become very well-known: five of our eight focus groups referred to it ('That's reminded me, they've banned conkers and stuff in schools [all agree]'[25]) and it was cited by nine of our 40 interviewees. The incident has been referenced by the UK Prime Minister[26] and the HSE included the story in its 'mythbusting' initiative.[27] A Google search for the terms 'conkers health and safety' produces more than one million hits.[28] There is, however, a catch: no rule, policy or incident compelled Halfpenny to make this decision. He has since referred to it as a light-hearted joke, a means of making an eye-catching point about risk aversion in the education sector.[29] Nevertheless the legacy of this 'regulatory myth' (Almond 2009) lives on.

Stories like this highlight the role of the media (and, increasingly, social media): regulatory myths depict a particular, extreme example as typical in order to make wider value statements about the legitimacy of health and safety regulation. The 'conkers' story was presented as saying something about British society: traditional activities and values of self-reliance were being eroded by the 'nanny state'. It continues to be used in this way by politicians and media outlets despite their *knowing* that it is inaccurate. It depicts the limits of any political consensus around health and safety: stories of this sort played a major part in bolstering the reform agenda of the post-2010 government (Young 2010). They are taken as evidence of 'public concern', despite their calculated origins.

This story, and others like it, demonstrate that the perceived legitimacy of health and safety regulation seems to diminish the further it extends into public life and 'low-risk' areas, as we noted with the DSE case. This breeds a resistance to health and safety which is a major problem for those, like local authorities, who manage risks: 'Every time I heard "conkers bonkers" [...] I knew that the rug was going to be pulled out from under my feet'.[30] But it also demonstrates that these tensions and determinations are *manufactured* by the media and used to advance particular regulatory preferences.

## The changing dimensions of legitimacy

On the basis of these case studies, and a wider review of our data, we identified 14 themes which seem to have contributed to shaping perceptions of the way in which the professions and organizations associated with health and safety in the UK have carried out their social licence to operate.

The first four themes relate to issues of '*constitutional*' legitimacy; the legal and institutional basis on which 'health and safety' is founded. The creation of *a single, unified inspectorate*, when the HSE was formed in 1974, provided an impetus towards greater consistency and a higher profile for health and safety. However, variation and conflict between specialist inspectorates continued and, arguably, some independence and expertise was lost. The larger HSE became, the greater the degree of political scrutiny (around accountability, budgets and decision-making). This led to the hiving off of the Railway, and Nuclear Installations, Inspectorates, and the abolition (in 2008) of the Health and Safety Commission, which was meant to shield the Executive from political pressure. It remains to be seen whether the HSE can retain its broadly constructive relations with political constituencies. *Government accountability* has also proved a challenge; HSC/E were created as non-Departmental bodies, designed to have functional and political independence from central government. However, as in other areas, the 'New Public Management' (Hood 1991) has emphasized accountability and control in the public sector leading to increased political intervention in shaping 'health and safety' as a policy area. This has conflicted with the arms-length nature of those bodies, meaning more reviews, more criticism and a more adversarial relationship. *The new legal framework* introduced by the HSWA 1974 imposed general duties, giving greater flexibility to duty-holders and more impetus towards self-regulation. However, firms have often seen flexibility as uncertainty and pressed for prescriptiveness at lower levels of implementation (regulations, Approved Codes of Practice, guidance notes, company policies and local procedures). This suggests that prescription may be an inevitable feature of the health and safety landscape, and responsible for some of the poor decisions and over-restrictiveness that underpin the negative media coverage of health and safety. Finally, *the influence of Europe* has been mixed. The initial 'creep' of EU competencies around health and safety, which peaked with the Framework Directive and 'six-pack regulations' (introduced into UK law in 1992) has plateaued. There is evidence that the UK was more active in shaping these measures than is often thought. Crucially, the public do not seem to perceive the role of the EU in health and safety to be a major problem – it concerns politicians and policymakers more than anyone else.

The second area of legitimacy challenge is '*democratic*', relating to the issues and groups that are addressed and represented within 'health and safety'. The *decline of traditional industries* means that the strongholds of industrial health and safety have fallen away over the last 50 years. Traditional 'leader industries' like mining, shipbuilding and textile manufacture have declined. This may mean that health and safety appears less vital to a larger proportion of the population than it once did. It also means that new ways of applying safety management principles in new settings (construction, retail and service sectors) are needed: health and safety remains an important issue in the new economy, but its legitimacy is less well-established. There has been a parallel *change in the nature of trade unionism*, including rates of membership and levels of influence over government. This is partly because the changes in industry have eroded traditional unionized sectors. However, successive governments over the last 30 years (Harvey 2005) have chosen to constrain the role of trade unions. This has undermined the model of tripartism – partnership between employers, unions and the state – which strongly influenced the original design of the UK health and safety regime. Nevertheless, there are still signs of an enduring influence, especially around worker engagement with safety issues at a local level. Lastly, the status of '*health' as the forgotten element* of health and safety has recently started to shift. Having been relatively downplayed for many years (because it is more diffuse, less visible and less easily addressed than 'safety'), health concerns have increasingly come to the fore, particularly in new workplaces (around musculoskeletal disorder, stress and psychosocial risks). Addressing these risks invites legitimacy challenges because interventions can appear more wide-ranging, precautionary and less oriented to immediately obvious and visible hazards.

The third set of crosscutting themes relate to '*functional*' legitimacy, around how 'health and safety' works in practice. First, the scope of health and safety has extended *beyond the workplace*, both in terms of risks caused by industry to the public 'over the factory wall' and into areas of private and social life. Section 3 of the HSWA prompted a shift towards taking account of 'public' risk. The parameters of this 'social safety' agenda have been heavily contested: there is some evidence that it has prompted the public and media backlash against health and safety in recent years (Almond 2009; Dunlop 2014). Second, the emergence and *development of a 'safety profession'*, of people and organizations whose role revolves around risk management and prevention, has created a sense of identity and interest around health and safety. Safety professionals have assumed this role partly because they are valuable (particularly since written risk assessments were introduced), and partly because they appear to be a 'preferred negotiating partner' for other groups. Their real weaknesses in terms of perceived legitimacy relate to a sense that there is a self-interested 'compliance industry' which distracts from the pursuit of outcomes that are not in the common interest, and a difficulty in communicating effectively with others. The great strength of 'health and safety' is the *perceived expertise* that underpins it: public perceptions do not suggest that a lack of expertise (on the part of regulators, safety professionals or others) is seen as a problem. In fact, this expertise seems to be widely valued. The challenges that arise here are about developing new expertise in emergent or contested areas, and the sharing of expertise with others in an effective way, so that competence and understanding is diffused and does not become a barrier (via technocracy and jargon: Black 2008). Lastly, the *evolving practices and profile of regulators*, particularly around enforcement and targeting, has an effect in terms of legitimacy; while increased efficiency has been secured, this has come at the cost of moving to an increasingly reactive mode (with even proactive inspections being in response to risk indicators). Arguably, this has limited dialogue and distanced regulators from the populations they interact with, and impacted on the societal appreciation and understanding of health and safety.

The final crosscutting themes relate to *'justice-based'* legitimacy, or questions about what moral values are being pursued, and the rightness of the outcomes produced. There has been a movement towards the *commercialization* of health and safety provision: consultancy and safety providers have become a visible part of the safety profession; regulators are seeking to recover costs (via 'fee for intervention', etc.); and the insurance and legal services industries are increasingly visible and active. In terms of public perception and legitimacy, this commercialism was viewed very negatively. It seemed to distort the altruistic reasons for action that were associated with health and safety as a public good. Commercialization may increase the reach and penetration of health and safety interventions, but many people do not appear to be convinced of the reasons why it might be preferable to a traditional public benefit model. The *limits of the consensus* around health and safety, which was the basis of the HSWA's tripartite approach, have been exposed, although this consensus was never as complete as was claimed in the 1970s. Health and safety interventions have always been contested, particularly where safety and profitability are in tension with one another. In recent years, though, this contest has become more explicitly political. Health and safety has once more become a subject for 'high politics', debates around the very principle and rightness of regulation, rather than the relatively stable 'low politics' of implementation (Moran 2003). Lastly, notions of *autonomy, choice and masculinity* continue to influence perceptions of health and safety, limiting the conditions under, and extent to which people are willing to engage with health and safety as a feature of modern life. The ideas of voluntary assumption of risk (that individuals, not the state, should decide the risks they are willing to undertake) and the notion of 'bargaining' (determining the balance of risk versus reward/payment that is drawn up) continue to shape behaviour and attitudes towards health and safety as they did in the nineteenth century. Gender roles, as traditionally constituted, continue to shape behaviours around safety and the assumption of risk, either via a 'masculinized' pressure to take pride in accepting risks, or a 'feminized' obligation to place caring for others ahead of one's own health and well-being.

## Influencing legitimacy: eight key insights

What can the professions and organizations involved with health and safety learn from this examination of their problems of legitimacy? We have identified eight elements that should be considered in developing any strategic responses.

### *Signal and noise*

Several of the case studies, particularly Flixborough, Nuclear and DSE, show that engagement with the public is important for everyone working in health and safety. Public perceptions of health and safety have two tiers: *opinions* (immediate, critical, transient), and *attitudes* (considered, positive, enduring). We can think of the difference between these as the difference between the sorts of statement someone might laugh at or agree with, and the propositions that they might vote for. When people stop to think about what health and safety means, their attitudes are much less critical. One key challenge, then, is to consider *how* to get people to consider health and safety more fully. Those engaged in 'selling' health and safety must separate 'signal' from 'noise'. It is not essential to leap in and contest every regulatory myth. On the other hand, there is a need to show that public concerns are taken seriously, preferably through proactive, upstream engagement. This means developing interactive dialogues – both talking and listening – as new technologies emerge or new production facilities are planned. It is not sufficient to present a modern public audience with an announcement of what has been decided and an invitation to comment. Health and safety also needs to be promoted through 'good news' stories, which would have some of the personal, emotional and dramatic features of regulatory myths, without stepping over the line into scaremongering.

### Share expertise

Expertise is a key asset of the health and safety regime – but like many social assets, it tends to be distributed very unevenly. As the Asbestosis case study showed, there would be benefits from making this expertise more widely available (as Löfstedt 2011a identified), improving the ability of a wider population of workers and the public to understand, assess and respond to, risks. While it will still be important to sustain and protect sources of specialist knowledge, basic competence in making risk and safety decisions should be more widely shared. This competence needs to be embedded in the generalist training of a range of people in managerial, supervisory and planning roles. At the same time, as later chapters examine in more detail, safety professionals need to move beyond hierarchical, command and control models of compliance to recognize and incorporate the knowledge and creativity of frontline workers. They can be a source of solutions as well as problems. One interviewee gave an example:

> you need to wear eye protection on a construction site […] we say that you don't need to wear eye protection if you're working outdoors and it's raining, because […] it gets covered in rainwater and makes it harder for you to see […] safety practitioners need to be alive to this and allow more choice, more flexibility.[31]

This approach should lead on to greater buy-in and engagement by frontline workers and more effective self-protection around hazards like asbestos. More generally, the role of safety professionals is enhanced rather than undermined by being seen as people who are willing to share their expertise in a process of dialogue and collaborative problem-solving rather than just declaring how things must be done. This also preserves credibility for those occasions in high-risk environments where strict adherence to safety protocols is essential.

### Establish opportunities for public debate

Beyond sharing expertise comes determination to involve a wider range of voices in the process of setting, directing and evaluating health and safety. The Nuclear case study shows vividly the need for, and potential benefits from, this kind of engagement. Writers such as Moore (1995) suggest that 'public value' should be seen as the objective of regulatory processes – the pursuit of goals and outcomes that are of value to society as a whole. Governing for public value involves creating opportunities for debating competing values and interests. These need to be open to people and groups beyond self-identified experts. Essentially, this means talking more, and more widely, to reach a mutual, if not mutually agreed, understanding about what government can and should try to do. The 'red tape challenge' of 2011–12 was a limited example of such a process. Wider engagement supports a more positive framing of 'health and safety' as something that everyone can be concerned about and participate in. The health and safety professions and organizations may be able to play a key role in offering 'core mediating spaces' where regulators, employers, workers and the public can meet on equal terms on neutral territory.

### Promote role engagement

Structured worker engagement with health and safety is a major component of efforts to legitimize and improve health and safety provision (James and Walters 2002; Walters 2006). Nevertheless, despite the advances documented in the Safety Representatives case study, there is a problem with the translation of this model into sectors and workplaces where union representation and activity is less well developed, or, more generally, into a world where union membership as a whole is lower than in the past. This is borne out by the figures on the level of worker representation in the UK: 95 per cent of workers are aware of health and safety (high), but only 66 per cent are consulted about it (low) (Jain and Leka 2015: 25–28). Worker engagement with health and safety needs to be pursued via adequate, active and genuine modes of consultation and representation. Safety representative roles must be valued if they are to be filled effectively. This would enhance both the democratic and functional legitimacy of health and safety.

### A 'right to safety'?

How can the *problem* of health and safety achieve a higher profile? Increasing knowledge of the extent of the problem would make it easier to justify regulation or intervention. In the past, as the Piper Alpha case study shows, major incidents and disasters have brought the problem to public attention. In recent years, we have enjoyed a relatively disaster-free period, compared with some previous times. No one wants another Piper Alpha to expose the consequences of neglecting workers' rights to safety. Some commentators have suggested that (unlike environmental regulation, which sits under a wider 'green' agenda) 'health and safety' lacks an appealing or easily understood overarching public narrative (Moran 2003). While this is a long-standing problem, the development of such a narrative would be helpful. Ideas such as that of a '*right to safety*', have the potential to work as powerful tools for framing public expectations. Shouldn't every worker be able to expect that their employment will not normally be the cause of death, disability or serious health impairment? Shouldn't everyone who lives next to a factory be able to expect that this will not kill them?

### Encourage questioning

Existing regulatory 'challenge panels', which allow people to question the use of 'health and safety' as a justification for action or in-action, are rather top-down. But, as the DSE case study shows, there is also a need for localized, bottom-up approaches. Can we encourage individuals to ask more questions face-to-face when they encounter health and safety in practice? How can we prevent the 'closing down' of conversations by manufacturers, service providers or public authorities who are using health and safety as an excuse for decisions that are really made on grounds of cost or convenience? If health and safety is to develop through explanation and engagement, there must also be encouragement for individuals to question decisions, without fearing retribution for doing so. If a decision has a sound justification, explaining this will build trust and confidence in the decision maker.

### *Be realistic*

Many of the events in the case studies provoked intense public, political and media scrutiny of health and safety. While health and safety professionals might wish for a world in which their activities are presented more positively, it is important to be realistic and pragmatic about what is achievable in this area. For example, there is a strong, positive story to be told about the role of trades unions in the modern world of work, taking more bottom-up, issues-led approaches to problems that concern them, their members and responsible employers. This contrasts with their current capacity, limited by the actions of successive governments, to influence the top-down, 'high politics' of health and safety. Pragmatism (including, for example, the recognition of the power of small-scale, localized change) is a valuable element in making health and safety work, particularly against a backdrop of media hostility and policy conflict at the level of 'high politics'.

### *Deal with prescription and variation*

The Conkers in the Playground case study nicely illustrates the recurrent problem of prescription. Robens (1972) sought to move health and safety away from prescription towards more flexible forms of regulation based on duties and principles. This was reflected in the HSWA and has continued to be an objective. However, subsequent experiences – with new industries and low-risk workplaces, as in the DSE case; with the public sphere of safety, as in the Conkers case; and with the less transparent fields of insurance, litigation and self-regulation – suggest that prescriptiveness has not gone away. Like the air in a blow-up mattress, prescriptiveness always seems to pop up somewhere in the system, despite efforts to squeeze it out. 'Blue tape' – requirements imposed by other commercial actors[32] – is an emergent issue, as is the tendency for organizations (like schools) to self-regulate via prescriptive internal rules and policies. Regulators and policymakers need to be aware that this occurs and appears to be inevitable. Rather than weighing in against it, their resources may be better used in working with safety professionals and duty-holders to create 'sensible' prescriptiveness, meeting the needs demonstrated by the drift towards prescription, but limiting the associated costs.

## Conclusion

This project has shown that the concept of 'health and safety' has had a complex history which encompasses many different trends, events and themes. Much has changed in this area since 1960 at political, contextual and institutional levels. Many of these interrelated changes have had significant implications for the ways in which health and safety is perceived and understood by different audiences. One key finding is that, while public perceptions of health and safety have remained more stable, and more positive, than might be thought, there is evidence that political and media discussion of health and safety has become increasingly polarized, particularly in the last 20 years. While the case studies here are necessarily selective in nature, they illustrate core challenges for those who wish to advance the cause of health and safety – around clear public communication, the establishment of strong narratives about the moral importance of health and safety, establishing new forums for the promotion of health and safety issues and a greater willingness to embrace the need for flexibility and debate. Participation and engagement at the level of individual employees, workplaces and citizens are vital if the tendency of the health and safety system towards insularity and the growing problems of political legitimacy are to be offset.

## Key points

- All institutions, organizations and professions need to be concerned with being perceived as legitimate by the society within which they are operating. If they are not thought to behave in a rational, reasonable and proper fashion, other parties are less willing to work with them or demand a higher price for doing so.
- The maintenance of legitimacy requires adaptation to social, cultural and technological change. The legitimacy of health and safety interventions has always been unstable because of the way in which it potentially conflicts with other social values like personal autonomy in risk taking and the freedom of private enterprises to maximize profit.
- Occupational health and safety has been much affected by its association with the impact of the European Union on domestic law and regulation; by changes in the industrial structure; by the broadening concern with less tangible aspects of employee health and community safety; and by the rise of a health and safety profession with its own commercial interests.
- The health and safety profession needs a new strategy for engaging a wider range of social partners in support of the objectives of occupational health and safety.

## Notes

1 Janet Asherson interview, para.4.
2 Cyril Bell, Flixborough plant safety manager, interviewed for BBC TV programme 'Red Alert', aired on 5 July 1977. BBC Written Archives, Caversham.

3 Anonymous interviewee ('SS'), former superintendent stevedore, Glasgow docks, discussing the 1960s. Interview by David Walker, 2009, Glasgow Dock Workers Project, Scottish Oral History Centre Archives (thereafter SOHCA) and CSG CIC Glasgow Museums and Libraries Collections.
4 David Eves interview, para.58.
5 Anonymous interviewee ('SS') 2009.
6 David Morris interview, para.28; anonymous interviewee ('SS') 2009; Owen McIntyre, former dock worker, Glasgow docks, discussing the 1960s and 1970s. Interview by David Walker, 2009, Glasgow Dock Workers Project, SOHCA and CSG CIC Glasgow Museums and Libraries Collections.
7 Dan Shears interview, para.94.
8 CORE/Greenpeace, 'Behind closed doors. Malpractice and incidents at Sellafield' (1988); HSE press release, 'HSE investigation of allegations of safety malpractices and incidents at BNFL Sellafield' (3 October 1988). British Safety Council archive, Mansfield, Pallet 6, box 35.
9 Paul Thomas interview, para.53.
10 TUC, 'TUC Representatives' Comments, Industrial Safety Advisory Council, Joint Safety Organisation' (10 July 1968): 1. Modern Records Centre, University of Warwick (hereafter MRC), MSS.292B/146.17.2.
11 David Eves interview, para.21.
12 Janet Asherson interview, para.45.
13 Minutes of Shop Stewards' Committee meeting, Reynold Chains Ltd, Coventry Chain Works (5 March 1980), minute 24. MRC, MSS.249/1/24.
14 Frank Doran MP interview, para.19.
15 Peter Jacques interview, p.30.
16 David Maidment interview, para.228.
17 Written testimony of the National Federation of Professional Workers to the Robens Committee, 15 March 1971, p. 2. The National Archives of the UK, London, LAB 96/967.
18 See 'Report from International Metalworkers Federation', in correspondence of the Iron and Steel Trades Confederation's national officer for health and safety, 1982, pp. 30–31. MRC MSS.36.2000.52.
19 Lawrence Waterman interview, para.64.
20 David Morris interview, para.111.
21 Jenny Bacon interview, para.63.
22 Neal Stone interview, para.57.
23 Focus group participants E8 and E7, 21.15–20.
24 http://news.bbc.co.uk/1/hi/england/cumbria/3712764.stm
25 Focus Group participant B3: 26.08.
26 '[W]hen children are made to wear goggles by their headteacher to play conkers ... what began as a noble intention to protect people from harm has mutated into a stultifying blanket of bureaucracy'. David Cameron, Speech to Policy. Exchange,<http://news.bbc.co.uk/1/hi/uk_politics/8388025.stm>, 01/12/2009.
27 www.hse.gov.uk/myth/september.htm
28 Search on 30 August 2016.
29 www.theguardian.com/commentisfree/2009/dec/09/conkers-goggles-myth-health-safety
30 Steve Sumner interview, para.63.
31 Lawrence Waterman interview, para.68.
32 www.telegraph.co.uk/news/politics/labour/10650716/Blue-tape-is-holding-business-back.html

# 2 The changing landscape of occupational safety and health policy in the UK

*Stavroula Leka, Aditya Jain, Gerard Zwetsloot, Nicholas Andreou and David Hollis*

## Introduction

The previous chapter discussed the changing social, economic and cultural contexts of health and safety policy and practice in the UK. Within the context of these changes, this chapter looks more closely at the policies themselves. It describes how the most important stakeholders have responded to changes in the world of work, and in societal expectations, through the development of regulation, including self-regulation, and standards. These developments are not unique to the UK. In recent years, many countries, in Europe and elsewhere, have sought to complement 'hard' policy in health and safety, such as legislation, with 'softer' forms of policy such as guidance and stakeholder initiatives (Leka et al. 2011). This new landscape reflects a number of contextual changes, including budget cuts, a lack of personnel and expertise within formal government bodies and an increasingly collaborative approach to policy development with business and other social partners (EU-OSHA 2009). We shall begin by reviewing the historical development of health and safety regulation in the UK and the current framework of legislation and policy. The chapter will then look at some of the factors that have been particularly influential in driving change and the initiatives that have resulted from these. Finally, we shall consider what wider lessons might be learned for the professions and organizations involved in occupational safety and health.

The material presented in this chapter is drawn upon the findings of a literature and policy review and three qualitative studies based on case studies, interviews and workshops with key stakeholders in the field of occupational safety and health. A number of sources were used for the policy and literature review including journal articles, books and dissertations; publications and reports of various interest groups; government publications and research documents; and the Internet. The search used relevant keywords and was broken down covering both the academic and grey literature. Studies were selected that had originated from credible sources (published journal articles, books, stakeholder websites and reports, and practitioner journals). The reference lists of any relevant source were further searched for subsequent references. Stakeholders identified as key actors in the health and safety arena throughout the research project participated in the subsequent qualitative studies. As wide a representation of stakeholders as possible was sought. In total, our research included a case study analysis of 15 policy initiatives in occupational health and safety, 40 stakeholder interviews, and two stakeholder workshops and focus groups with about ten participants in each.

## Historical overview of occupational health and safety policy in the UK

According to the *Joint International Labour Organization and World Health Organization Committee on Occupational Health*, occupational health and safety should aim at:

> the promotion and maintenance of the highest degree of physical, mental and social well-being of workers in all occupations; the prevention amongst workers of departures from health caused by their working conditions; the protection of workers in their employment from risks resulting from factors adverse to health; the placing and maintenance of the worker in an occupational environment adapted to his physiological and psychological capabilities; and, to summarize, the adaptation of work to man and of each man to his job.
>
> (ILO/WHO 1950)

Health and safety regulation is not a new societal concern (Henshaw et al. 2007) but it has evolved significantly since its early days. Originating from the law of King Henry I (1068–1135), masters were responsible and liable for injuries to their servants, or loss of life, due to negligence (Rabinowitz 2002). In the small work units that existed before the industrial revolution, masters, or as we would now say, employers, often had greater knowledge of the tasks being performed by their servants, or workers. Because they were more knowledgeable about the risks, they should also have been accountable for preventing or minimizing them.

This model changed dramatically during the nineteenth century. With the industrial revolution, factories and mills emerged in swathes across the midlands and northern England, bringing with them large-scale activities, increased mechanization and increasingly dangerous working conditions. These developments led to frequent industrial accidents (Henshaw et al. 2007). Although industrial restructuring during this period outpaced the law's response, it did bring questions of occupational health and safety before Parliament and the courts (Barrett and Howells 1997). Those institutions acted through the only means at their disposal: legislation. This early landscape evolved in several directions as the world of work changed. Consequently, different approaches were implemented to advance health and safety standards, underpinned by different perspectives from diverse stakeholders. Figure 2.1 presents a timeline of critical events.

## The early years of health and safety regulation in the UK

The first occupational health and safety statute in the UK was introduced in 1802 by Sir Robert Peel. The *Health and Morals of Apprentices Act* targeted those employed in cotton mills and other factories. It restricted the working day to 12 hours and anticipated a phased elimination of night work (Callaghan 2007). This was followed by the *Factory Regulation Act* 1833, later amended through the *Factories Amendment Act* 1844. There was some apprehension about prosecuting employers for health and safety issues, because the jobs they provided were seen as a benefit to society, and no mechanism for enforcement was established (Johnstone and Carson 2002).

Subsequent legislation developed in reaction to specific situations or events: major accidents in coal mines, the identification of occupational diseases or the emergence of new hazards, like electricity, as a result of technological developments (Rimington et al. 2003). Specific regulations and enforcement bodies were created as each hazard was identified (e.g. the *Mining Inspectorate*, the *Railway Inspectorate* and the *Alkali Inspectorate*). This appears to be the origins of a hazard-based approach founded on the 'precautionary principle' (Löfstedt 2004; Rimington et al. 2003). For much of the nineteenth century, factory regulation only affected a minority of those working in manufacturing industry. Following a Royal Commission investigation, a consolidating measure, the *Factory and Workshop Act* 1878, was passed which, with later amendment in 1901, gave Ministers powers to pass *ad hoc* regulations to deal with hazards in certain industries (Callaghan 2007). In 1931, the *Asbestos Industry Regulations* were enacted, recognizing health-related issues. The *National Insurance Act* 1946 provided state benefits to those who were sick or unemployed, and insured all workers who had paid the necessary contributions against workplace accidents, while the *Shops Act* 1950 limited working hours for those under the age of 18 (Lyddon 2012).

By the 1960s this piecemeal approach to occupational health and safety had resulted in over 500 pieces of legislation (Fairman 1994). The *Factories Act* 1961 attempted to update and consolidate these, providing an overarching and comprehensive framework applicable to all types of industry. The reach of health and safety regulation was extended by the *Offices Act* 1960 and the *Offices, Shops and Railway Premises Act* 1963 (Lyddon 2010). Millions of employees in the service sector were given some degree of health and safety protection for the first time. The *Employers' Liability (Compulsory Insurance) Act* 1969 reflected a belief that economic incentives were also important in securing compliance by employers. This Act required most employers to take out employer's liability insurance to cover the cost of compensating employees or former employees for workplace injuries (HSE 2008).

Before 1974, UK health and safety legislation was characterized by a limited scope, patchy coverage of the workforce and slow adaptation to change (Lyddon 2010). The economy was dominated by a large industrial sector, whose workforce was highly unionized and politically active: in 1970, trade

*Figure 2.1* Timeline of critical events in the changing OSH landscape in the UK – 1802 to 2013

union membership reached 10.6 million, almost half of all salaried employees (Beck and Woolfson 2000). Despite the reforms of the early 1960s, the legislative framework was thought to be excessively detailed and over-complicated. Existing Acts were accompanied by hundreds of Statutory Instruments with a multitude of agencies overseeing the system (Crombie 2000). Despite the mass of regulation, the number of workplace accidents was not falling. A comprehensive review of the system was launched in 1970 by a committee chaired by Lord Robens, Chairman of the National Coal Board, who was a former trade union official and Labour politician (Rimington et al. 2003).

## The Health and Safety at Work etc. Act 1974

When the Robens Committee reported in 1972, it recommended a radical change in attitudes to health and safety on the part of both industry and government (Rimington et al. 2003). Over 150 years after the first health and safety legislation, the UK passed the *Health and Safety at Work etc. Act* (HSWA) 1974 in an attempt to rectify many of the flaws of the existing legislation. The passage of the Act coincided with the major explosive accident at the Nypro chemical plant in Flixborough, which was discussed in the previous chapter. The Act also coincided with the emergence of interest in occupational health and safety at the level of the European Community (Rimington et al. 2003). The HSWA created a framework for the general application of health and safety provisions to every workplace in Great Britain (Barrett and Howells 1997). It established broad rules but then allowed for the creation of more specific regulations and Approved Codes of Practice (ACoPs) where these were thought to be necessary. Although ACoPs had previously existed in certain sectors, the Act considerably extended their application. ACoPs do not impose legal duties: failing to comply with a provision of an AcoP is not in itself an offence. However, if a statutory requirement is not satisfied in some other way, then an employer could be found guilty of contravening the relevant regulation or section(s) of the HSWA (Fairman 1994).

The revised legislation, and any subsequent additions, were meant to set objectives rather than to prescribe the means by which these were to be achieved, focussing attention on prevention rather than compensation or retribution (Barrett and Howells 1997). This shift in policy was seen as a way to rectify the issue of 'apathy' that had been identified by the Robens committee, ensuring risk was dealt with by those who created it (Dalton 1998). Importantly, the general duties under the Act are qualified by the phrase 'so far as is reasonably practicable'. This acknowledges that there will always be a trade-off between safety and factors such as cost and feasibility (Fairman 1994). This philosophy can be seen as the origins of a risk-based approach to health and safety, although other authors have suggested that this only appeared in regulation during the 1980s (Hutter 2005). Carrying out what is reasonably practicable requires an understanding of the necessary risks in order to weigh up the cost of action relative to inaction (Bearfield 2009). Indeed, Callaghan (2007) notes that, through introducing the concept of risk, Robens anticipated the 'better regulation' agendas of later governments.

The 1974 Act provided for the establishment of the *Health and Safety Commission* and *Health and Safety Executive*. The Act also significantly expanded the sanctions available to both the HSE and local authority inspectors (Lyddon 2012). It allowed both parties to issue prohibition notices (where a dangerous activity had to stop until it had been made safe) and improvement notices (where an employer had a set time to comply). Three years after the HSWA Act, the *Safety Representatives and Safety Committees Regulations* (SRSC) were passed (Goldman and Lewis 2004). These regulations arose through the bargain struck between the Labour government and organized labour during 1974–79, referred to as the Social Contract (Eurofound 2009). They extended the duty described in the HSWA for employers to provide their employees with information necessary to ensure health and safety at work; and for employees in union-recognized firms to appoint safety representatives who could call for the establishment of safety committees, investigate potential hazards and complaints, and access information about health and safety matters. These representatives were entitled to paid leave to carry out these tasks (Robinson and Smallman 2006).

The HSWA looked to reduce the inefficiencies of health and safety, removing unnecessary elements while striving to increase the protection of workers, through shifting responsibility to employers (Aalders and Wilthagen 1997; Genn 1993). In many ways this can be seen as a deregulatory agenda, which was a radically new approach. Similarly, the 1977 SRSC regulations brought a culture change through the inclusion of workers. The period can be characterized by government (both Labour and Conservative) adopting a new approach to occupational health and safety, where employee protection was based on employee status rather than on the type of premises where the employee worked (Lyddon 2012).

### The turbulent Thatcher and post Thatcher period

The Thatcher government, established in 1979, brought new approaches to the governance of the UK. This model, which became known as 'Thatcherism', sought to reduce the role of the state to that of repairing market failures rather than actively managing economic activity (Tombs 1996). It was partly a response to the economic recession of the 1970s and early 1980s. Government efforts to free up markets were seen to be a way to promote business innovation and efficiency. Many policy areas with established regulation, such as occupational health and safety, witnessed attempted deregulation (Baggot 1989). In the mid 1980s, two White Papers were published, *Building Businesses Not Barriers* (Department of Employment 1986) and *Lifting the Burden* (Department of Environment 1985), both of which called for reductions in the general regulatory 'burden' on business. *Building Businesses Not Barriers* proposed, for the first time, a systematic review of regulations in order to identify where the 'burdens' lay. In practice, though, the UK's health and safety legislative and regulatory systems remained relatively untouched. The supposed chief beneficiaries of deregulation, small firms, did not seem to have found health and safety law particularly burdensome (Dalton 1991). Moreover, reform provoked considerable public concern, and opposition, including opposition from employers (Department of Environment 1985; Department of Employment 1986), forcing the government to take a more subtle approach to its deregulation agenda (Tombs 1996).

The *Employment Act* 1980 repealed provisions made under the *SRSC Regulations* and the *Employment Protection Act* 1975. In 1985, the *Reporting of Injuries, Diseases and Dangerous Occurrences Regulations* (RIDDOR) were introduced. These require a 'responsible person' to notify the enforcing authority in the event that: a fatality occurs; a person sustains any injuries or specific medical conditions; or where a dangerous occurrence takes place in connection with a work activity. Importantly, this reintroduced a requirement to report all incidents that led to more than three days of absence from the workplace (Dawson et al. 1988). The *Electricity at Work Regulations* and the *Noise at Work Regulations*, both introduced in 1989, illustrate the continuing uncertainty about how best to deal with health and safety issues: the former took a traditional prescriptive approach, while the latter were more goal-setting in nature.

The risk-oriented approach was also marked by the introduction of the doctrine of risk tolerability in the 1988 HSE document *The tolerability of risk from nuclear power stations*. This established a three-tiered approach to risk by the HSE, defining where organizations should implement health and safety initiatives. The approach is founded on the principle of 'as low as is reasonably practicable' (ALARP) rather than on the total elimination of all risk (Kemp 1991). The proportion of HSE expenditure which came directly from the government fell from 98 to 76 per cent between 1975/76 and 1990/91.

However, towards the end of the 1980s, several factors combined to press the government into giving occupational health and safety regulation a higher priority. First, there was an increase in 'major' injury rate for 18 of 19 industry groups between 1981 and 1985 (HSE 1985). Second, with the introduction of employee data to gauge working conditions, the 1990 Labour Force Survey (LFS) found that 1.5 million people reported work-related injuries in the previous year, 2 million people suffered from an illness that they believed to have been caused, or made worse, by work, and work-related injuries and ill health were linked to approximately 29 million lost working days annually (10 times more than those lost through strike action; Dalton 1992). Third, and perhaps most importantly, a number of highly publicized private and public disasters had led to fatalities and injuries, including: the *Herald of Free Enterprise* sinking; the Clapham rail crash; the Kings Cross fire; major incidents at the Hillsborough and Bradford football stadiums; Flixborough and Piper Alpha. These disasters, some of which were discussed in the previous chapter, focussed attention on the issues of risk management and commitment by top level management to health and safety issues.

### The increasing influence of the European Union

This period saw the development of a critical influence on the UK's occupational health and safety system through European Union (EU) regulation. Until 1987, the introduction of EU health and safety law into the UK was slow and piecemeal (Dalton 1998). This changed with the *Single European Act* 1987 and a move towards Directives which specified general duties and minimum standards with the aim of harmonization across EU member states. The *Framework Directive 89/391/EEC*, on the introduction of measures to encourage improvements in the safety and health of workers at work, was adopted in June 1989. This was followed by a number of individual Directives (Barrett and Howells 1997). The *Framework Directive* is focussed on systematic risk assessment and internal competence. Its preamble states that health and safety at work is an objective which should not be subordinated to purely economic considerations (Dalton 1998). It establishes the principle that the employer has a duty to ensure the safety and health of workers in every aspect related to their work. Crucially the Directive broadened the scope of health and safety with the inclusion of issues such as the organization of work, and social relationships. Individual Directives were designed to supplement the *Framework Directive* by tailoring its principles to specific issues.

On 1 January 1993, the *Framework Directive* and five subsidiary Directives were implemented in the UK by six new regulations (the 'six-pack'). Other notable regulations that were also passed under EU influence were the *Control of Substances Hazardous to Health Regulations* (COSHH), implemented in 1989 (later updated in 2002). Various statutes pre-dating the 1974 Act were also revised, leading to the simplification of approximately 40 items of legislation (Facilities 1993).

EU legislation has also had an influence in the arena of worker involvement. In 1996, the *Health and Safety (Consultation with Employees) Regulations* were enacted to give protection to employees who are not trade union members, entitling them to elect their own health and safety representatives, and, through them, to consult with employers about health and safety issues (Goldman and Lewis 2004). These regulations stemmed from concerns that restricting representation to unionized workers or organizations conflicted with Article 11 of the Framework Directive (which required all workers to have such rights; James and Walters 1997). The *Management of Health and Safety at Work Regulations* 1999 (MHSW) are also relevant: they furthered the self-regulation agenda through a focus on health and safety management, rather than structural changes, and a risk-based approach (Dalton 1998).

In this period, the concept of social dialogue also increasingly influenced the development of occupational health and safety policy. Social partners obtained the right to be consulted by the European Commission on all initiatives and to negotiate and conclude framework agreements, which could be adopted as EU law. As a result, three framework agreements were signed by social partners at European level on: parental leave in 1995, part-time work in 1997 and fixed time work in 1999 (Branch 2005). These were then implemented as Directives and transposed into national legislation. The gradual creep of European law was noted in the *Davidson Review* (2006) which found that between 1984 and 1993 more than half the occupational health and safety regulations laid before Parliament originated in Europe. This would contribute to resistance against Europe in later years.

Paralleling these EU developments, the UK health and safety policy arena also developed substantially during this period. In October 1992, the government launched another 'deregulation initiative' aimed at removing unnecessary red tape from business. One key proposal was the introduction of impact assessments (weighing the costs and the benefits) for regulatory reforms which would affect business, with any such measures being signed off by Ministers. At the same time, the government brought forward the *Deregulating and Contracting Out Act* 1994, which gave the Secretary of State for Employment powers to repeal legislation, including health and safety (Bain 1997). Following a review by HSC, seven pieces of primary legislation and 100 sets of regulations, were removed, although EU-based legislation was unaffected (HSC 1994). In 1995, public funding for training union health and safety representatives, available since the SRSC Regulations in 1977, was ended (Bain 1997).

Growing discontent over the extent of European influence on national legislation was another pressure for change. In September 1994, the European Council of Ministers set up a review body to examine the broad field of social policy, including health and safety legislation. This 'Molitor group' called for the removal of burdensome and unnecessary legislation, including a review of health and safety Directives, a decrease in administrative expenditure for trade and industry and a strengthening of the competitive position of businesses within the EU (Plomp 2008). Despite the focus on reducing the 'burden' of health and safety, consideration of health matters continued to expand. Following the findings of Health Risk Reviews in the 1980s which highlighted the significance of work-related stress (Mackay et al. 2004), the HSE published guidance on this topic for employers in 1995 (Spiers 2003).

In keeping with the reduced emphasis on regulation, the HSE produced a policy statement in 1996 which promoted standards as a form of guidance in health and safety and stated its commitment to standard making where there was justification to do so and resources permitted. BSI Management Systems, a management systems registrar, collaborated with occupational health and safety experts from around the world to create the OSHAS 18001 series (O'Connell 2004). Softer forms of policy were beginning to develop momentum.

## The new millennium

The 2000s saw a number of health and safety strategies issued by the HSE and the government. In 2000, *Revitalising health and safety* was partly driven by the failure of the number of working days lost figures to fall over the previous decade. It introduced a target-based approach into the UK health and safety system for the first time. This strategy was incorporated into the HSC's 2004 *A Strategy for Workplace Health to 2010 and Beyond*. This strategy was not target based but called for a more effective self-regulatory system with clearer delineation of the roles of the HSC, HSE and local authorities. It also advocated a more evidence-based approach to the system and to living and working with risk. The HSC/E's *Simplification Plan* outlined the agency's progress in reducing the administrative burdens of occupational health and safety. It reported the discontinuation of over half of HSE forms, increased use of the website to list primary and secondary legislation and a simplified reporting regime for accidents and incidents (HSE 2006). A follow-up strategy was launched in 2009 called *The Health and Safety of Great Britain: Be Part of the Solution* which aimed to instil the responsibility of all stakeholders for managing occupational health and safety. The 2009 Anderson review estimated that uncertainty regarding how to comply with legislation cost over £880 million annually and recommended the simplification of legislation for easier interpretation by business.

There were also more specific initiatives in the 2000s. In order to meet the targets for reduction of occupational ill health and absence in the UK, the HSC identified tackling work-related stress as a priority (HSE [no date]). Instead of introducing an ACoP, Management Standards for work-related stress were adopted in 2004, based on an agreed framework of good management practice. The *Construction (Design and Management) Regulations* (CDM) were launched in 2007, while on 6 April 2008, the *Corporate Manslaughter and Corporate Homicide Act* 2007 came into force. With this Act, an organization will be guilty if the way it manages or organizes its activities causes a death and amounts to a gross breach of a duty of care to the deceased. The *Health and Safety (Offences) Act* 2008 came into force in 2009, increasing the fines for health and safety offences from £5,000 to £20,000 and extending the range of offences for which an individual (manager) can be imprisoned, strengthening accountability and economic incentives for managing health and safety.

Developments also occurred at the European level. The European Commission formed two strategies addressing the negative effects of occupational ill health. EU influence on UK health and safety legislation continued, with a significant addition being the *Registration, Evaluation, Authorisation and Restriction of Chemicals (REACH)* Regulation, which came into force in the UK in June 2007. This established a new approach to the control of chemicals by shifting responsibility for substance risk assessment from the regulator to manufacturers and importers. The 2000s also witnessed a culmination of the EU social partners' social dialogue efforts with the conclusion of three autonomous agreements on: tele-work, 2002; work-related stress, 2003; and harassment and violence at work, 2007. None of these were transposed into EU legislation (Leka et al. 2010).

Deregulatory agendas in the EU also advanced. The 2000 Lisbon European Council set a goal for the EU to become the most competitive and dynamic knowledge-based economy in the world by 2010. To achieve this, the Council asked various EU departments to use regulatory impact assessments for all proposed regulations. Neal (2004) suggests that the European position on occupational health and safety policy has since changed markedly towards a deregulatory agenda. These authors argue that, as suggested by Löfstedt (2004), there have been marked moves towards the reclassification and reordering of existing Directives, updating earlier Directives to ensure consistency with technical progress, and 'soft law' solutions.

### Recent developments in occupational health and safety in the UK

The 2010 coalition government commissioned two major reviews of occupational health and safety. In 2010, Lord Young's review, *Common Sense, Common Safety* on legislation and the (alleged) compensation culture surrounding health and safety was published. The 2011 Löfstedt review aimed to consider the opportunities for reducing the burden of health and safety legislation while maintaining the progress the UK has made in improving outcomes. These reviews took place in a wider deregulatory context, as the *Red Tape Challenge* initiative aimed to reduce the amount of legislation on the statute book. As part of this Challenge, the HSE was also involved in a Star Chamber Review, having to find cost savings of 35 per cent by 2014/15 (*Safety & Health Practitioner*, 2010).

The Young review (Young, 2010) found that businesses have issues with the interpretation, rather than the content, of legislation and that the compensation culture surrounding health and safety was 'perception rather than reality' (p.19). The review also proposed classifying workplaces according to their level of risk (although ignoring the impact of occupational health concerns in this classification and directly questioning the notion of evidence-based policymaking), and recommended that the UK 'takes the lead in cooperating with other member states to ensure EU health and safety rules for low risk businesses are not overly prescriptive, are proportionate, and do not attempt to eliminate all risk' (p.40). The review also recommended that the RIDDOR reporting period for employees absent from work following an accident or injury at work be extended from three to seven days; and that the level of professionalism for occupational health and safety consultants should be increased by requiring them to hold a qualification with a professional body. Following the latter recommendation, the *Occupational Safety and Health Consultants Register* (OSHCR) was set up in 2010.

In a similar vein, Löfstedt (2011a) found no evidence for radically altering health and safety legislation. He acknowledged that previous deregulatory actions had reduced health and safety regulation by 46 per cent, compared with 35 years previously. However, he did recommend the consolidation of sector-specific legislation by 35 per cent and a review of all 53 ACoPs with a view to simplification. Löfstedt made other recommendations that were contested by some stakeholders, such as the exemption from health and safety legislation of the self-employed and those working from home, where their work activities pose no potential harm to others, and the revision of the *Construction (Design and Management) Regulations* 2007 and their associated ACoP.

The guiding principle behind the Löfstedt review was that regulation should be risk-based rather than hazard-based. Löfstedt agreed with Lord Young that the concerns businesses have with health and safety are less with the regulations themselves, than with the way they are interpreted and applied. Inconsistent regulatory enforcement activity, and the negative influence of third parties, generates unnecessary paperwork and a demand for health and safety activities which goes beyond the regulatory requirements. The government's subsequent overreaching of the recommendations, and their proposed timescale for enacting changes, has been criticized by many stakeholders (Wustemann 2011). There has also been controversy over the HSE's newly introduced fee-for-intervention scheme where duty-holders who materially breach health and safety law will be charged £124 per hour for the HSE's intervention, in an extension of the cost recovery principle. This has shifted the role of HSE inspectors towards a consultancy model.

In reviewing the changing landscape of occupational health and safety regulation in the UK, it becomes apparent that history repeats itself in many ways. This would include the definition of priorities, such as reducing the burden on businesses, especially SMEs, and deregulation in all its forms, from reviewing and rationalizing legislation to budget cuts and weakening relevant government agencies. However, as we saw in the previous chapter, the health and safety landscape is not independent of wider influences, including social, economic and political. These not only define how health and safety are dealt with but also the nature of work itself. Stakeholders have often been slow to respond to changes in both the context and in working practices. The more complex the landscape becomes, in terms of the influences it receives (and their outcomes) and the actors that emerge, the more flexibility is required in the systems that control it, evidenced by the increasingly diverse forms of regulation implemented in recent years. Where health and safety regulation used to be prescriptive and rigid, it has evolved into being goal-setting and risk-based, with voluntary forms of regulation (soft law) added to legislation (hard law). More stakeholders than ever before are active in the health and safety landscape, promoting their own approaches to regulation. The question then becomes one of how best to balance these competing pressures and policy options in order to achieve desirable outcomes.

## The current landscape of occupational health and safety

The key message of the current national occupational health and safety strategy in the UK is that everyone has a role to play in improving the country's performance. Improvements will only be achieved by all stakeholders working together towards a set of common goals. If this is to become a reality, every stakeholder has to understand their own role and become better at executing their responsibilities. Key stakeholders include: employers and their representative bodies; the self-employed; workers and their representative bodies; government, through its departments and agencies (*Health and Safety Executive*, local authorities, etc.); professional bodies; voluntary and third sector organizations (HSE 2009). There is a complex network of sources of health and safety support, advice and information available to both employers and workers which are largely outside the direct control of either the HSE or local authorities which together form the national health and safety landscape in the UK (BRE 2008). This will be explored in more detail in the next chapter.

The diverse involvements (see Table 2.1) of stakeholders contribute to the dynamism of the health and safety policy process (Jain et al. 2011; Zwetsloot et al. 2008). This is evident in the types of policy initiatives they have promoted. They include a range of legislative (i.e. 'hard') and non-legislative (i.e. 'soft') policy-level interventions engaging a variety of stakeholders (for example, government, trade associations, employer associations, standardization bodies). Initiatives have included: legislation; development of national strategy and accident reduction schemes; standards and certification; guidance, classifications and specifications; codes of practice; stakeholder/collective agreements (social dialogue); awareness raising campaigns; economic incentives/programmes; networks/partnerships; performance evaluation; and benchmarking tools.

An analysis of 15 different types of initiative, promoted by various stakeholders, showed that key drivers for their development were legislation, regulatory reviews and meeting industry/market needs, while key challenges included resource availability, perception, measurement, and meeting demand. Seven elements appear to be essential for the success of such initiatives (see Table 2.2).

The perception of the initiative and the degree of engagement with it are relevant issues for both policymakers and the target audience. Where legislative reform is seen to have come from Europe, resistance has often been observed from various stakeholders such as UK policymakers themselves, businesses and the general public (Löfstedt 2011b). This may explain some of the perceptions of health and safety identified by Lord Young's (2010, p.5) observation that 'the standing of health and safety in the eyes of the public has never been lower'. As such, the initiatives that we reviewed were not acknowledged by the participants as political successes in terms of having a direct effect on the popularity of government. However, the popularity of some initiatives did enhance the reputation of the stakeholders who developed them.

*Table 2.1* Stakeholders in OSH management and their main stakes

| Stakeholders | Main stakes |
| --- | --- |
| Employers | Good OSH management is of primary importance to ensure that workers remain healthy and productive. Employers also have a legal obligation to provide safe and healthy workplaces. |
| Employees | Good OSH management is of primary importance to employees for their own health and productivity (staying economically active). They share the legal obligation with employers. |
| Government agencies | Develop and implement OSH regulation. Monitoring, inspection and ensuring compliance with national OSH regulations and standards. Provision of basic OSH services, for example through primary health care system. |
| Researchers and academics | Development of OSH management tools. Examine the link between exposure to occupational risks and health and share this information with practitioners. |
| OSH services | Implement OSH management initiatives and tools. |
| Social security agencies | Good OSH management may reduce the burden of disease and help to reduce rising costs of healthcare on social security arrangements (for workers compensation, societal costs of disabilities and associated unemployment). Social security agencies have a clear stake in prevention. |
| Health insurers | Good OSH management may reduce the rise of health care costs for treatment of occupational diseases. Health insurers have a clear stake in (primary and secondary) prevention. |
| Health care institutions | The prevalence of occupational health problems is a challenge and burden to the health care systems and institutions. Increasing treatment activities may trigger greater interest in prevention. |
| Customers/clients | In many jobs people work with clients. If workers suffer from illnesses, this is likely to affect the way they work which may also reduce customer satisfaction. |
| Shareholders | Occupational ill health can lead to high levels of sickness absence. In companies with severe problems, it may also be more difficult to attract talent. As a result the productivity and competitiveness of the company may be affected, implying reduced shareholder value. |
| NGOs/civil society | NGOs represent civil society groups. Several civil society groups may have an interest in good OSH management by companies. This may include expectations of social benefits to be generated as a result of the hosting of the business activities (e.g. access to employment, improved livelihoods). Potential change in communicable disease patterns (spread of diseases from workers to members of wider community or vice-versa). Threat of injury from violence (for example as a result of the inappropriate use and training of security personnel, poor practices in managing site safety, etc.). |
| Universities, business schools and vocational institutes | Good OSH management clearly has a link with good business practice. This is important for the education of present and future business leaders and workers. OSH management should therefore be integrated in the curricula of universities, business schools and vocational institutes. |

*Table 2.1 (cont.)*

| Stakeholders | Main stakes |
| --- | --- |
| Employment agencies | Occupational disorders (particularly psychosocial disorders) are increasingly relevant as a cause of reduced work ability and rising unemployment. Recent literature shows that (re)activation of long-term unemployed persons is more successful when it is combined with work than in the traditional model of treatment and cure before people start working. This implies that employment agencies have a clear interest in tertiary prevention. |
| Human resource departments and officers | Within companies, OSH issues are relevant for accident prevention, well-being at work, company climate, employee satisfaction, and the retention of existing employees. Though coming from another tradition compared to OSH experts, HR officers are increasingly involved in the management of OSH issues, particularly psychosocial issues, at work. |
| Media | OSH management is a societal issue with ever growing impact. It is important to many people (workers, their families, etc.). As a result the issue is of growing importance to mass media (journals, TV, Internet, etc.). |
| Actors of (in) the judiciary system | OSH risks are increasingly having economic implications both for companies and their workers. This is likely to lead to a boost in legal cases on liability issues. This may form a burden to parts of the legal system but might be a source of potential income to lawyers. |
| Business consultants | As OSH risks are increasingly having business impacts, advising on these issues will probably not remain the exclusive domain of occupational health and safety services. Business consultants are likely to develop a growing interest in this area. |

*Source: Adapted from Zwetsloot et al. 2008*

These conclusions echo Baril-Gingras et al.'s (2006) findings on organizational level interventions that showed managers and supervisors to have a strong influence on whether changes were executed or not. This concurs with the views of several policy developers within the current study. Senior management within the HSE were, for example, identified as crucial gatekeepers at the policy development stage, at the policy implementation stage via the scheme champions (who acted as the change agents) and among the target group, the trade associations and the companies within the four sectors. Across all of these actors, managers were critical to turning intellectual commitment or 'buy-in' into enacted behavioural change. Political will and resources were identified as key success facilitators within the present study. At the same time several respondents noted the emphasis on deregulation and the use of health and safety as a 'political football' as evidence that the perception of this as burdensome was driven by political ideology rather than evidence.

Various drivers motivate stakeholders to adhere to best practice. Respondents identified three key pillars: legal, moral/corporate social responsibility, and

*Table 2.2* Key elements for OSH policy initiative success

*Link to identified need*: The policy initiative has to meet an identified need in an appropriate manner.

*Legitimacy:* The policy initiative has to have legitimacy in meeting an identified need. This might relate to it being promoted by an authority such as the European Commission or experts, or having a strong evidence base to support its implementation.

*Clear ownership and commitment:* Ownership of, and commitment to, the initiative by recognized stakeholders (such as the government, social partners, trade bodies or sectoral bodies) is essential.

*Consultation:* Consultation with various stakeholders and raising awareness in relation to the initiative is important. This consultation should happen in a structured, systematic and transparent process. The initiative will have a greater chance of success if there is the right balance between different stakeholder interests depending on social, economic and political influences.

*Resource availability:* The continuity of a policy initiative is dependent on its resource availability in terms of finances, personnel and time.

*Structured implementation:* Successful initiatives are implemented through a structured process including clear objectives, responsibilities.

*Policy evaluation:* Evaluation methods that allow learning, knowledge transfer and future initiative development are critical for the success of an initiative.

financial. These findings broadly echo previous research (e.g. Wright et al. 2005; Wright 1998; Wright et al. 2004) that has found the main drivers to be: enforcement/regulation; reputational risk; the moral case; avoiding cost of accidents; business incentives; and supply chain pressures.

Meeting the legal requirements of UK law and fulfilling Member obligations to EU law were found to be co-occurring drivers for multiple initiatives. Culyer et al. (2008) note a plethora of reasons why workplace health and safety regulation and minimum standards are required through government intervention. Due to the absence of full employment, imperfect information, lack of perfect labour mobility across jobs and the existence of externalities, the market, left to its own devices, will not yield an optimal level of compensation for risk. Workers would therefore face a higher than optimal level of risk. While Kankaanpää and colleagues (2008) note that, in principle, a competitive market is supposed to lead to the optimal allocation of resource, they acknowledge that occupational health and safety is, in this respect, an anomaly. The authors note that, in developed countries (such as the UK), employers are not always required to handle the negative externalities of production – the costs of ill health and disability – by themselves. This is because the risks of occupational accidents and diseases have been pooled, so the costs of lower than optimal health and safety are covered by society. However, the market emerged in our study as a key driver for high standards, alongside regulatory influences and reviews. The 'business case' for health and safety was identified in Wright and Marsden's (2005) analysis of survey data from 500 organizations. They found 73 per cent of employers believed health and safety requirements benefited their business as a whole, while 64 per cent believed they saved money in the long term because of them.

It was evident from our respondents that the cost of insurance was a key factor in motivating firms (specifically board members) to take their commitments seriously. Wright et al. (2005) stated that, while insurance was not an important driver for firms during the 1990s, its importance had grown significantly from 2002 onwards. Indeed they found that, due to increases in cost, the role of insurance, specifically employer's liability insurance, was taking on increasing significance. Similar findings have been reported in European studies, where the business case has been highlighted as particularly important for the management of new types of health and safety risks, such as psychosocial risks, across all EU member states (EU-OSHA 2012).

Wright (1998) notes that so-called 'softer' drivers work with highly motivated, higher-risk firms. Companies operating in major hazard industries implement health and safety management measures as a matter of principle as much as for potential business benefits. The softer 'market-led' initiatives could also be considered in light of Budworth and Khan's (2000) continuous improvement model. This categorizes organizations as: those not interested in health and safety issues; the compliers; and the advocates. The authors note the first group are likely to be driven by enforcement and regulation, whilst the last group are influenced by reputation and longer-term costs.

We found that the moral imperative of corporate social responsibility was the second most important driver behind legal compliance. It should be considered as a supplement to the 'stick' approach. Research has already illustrated that several companies have begun to cite health and safety-related interventions in corporate social responsibility (CSR) reports (Andreou and Leka 2013). The publication of guidance affects corporate practice, with CSR reporting converging on what is included in official or industry-wide standards (Chen and Bouvain 2009). Some governments, such as Denmark, have made publication of CSR-related communication a legal requirement in the hope that it would stimulate companies to move towards existing best practice (Danish Commerce and Companies Agency 2011). The UK and/or the EU could explore this avenue (Zandvliet 2011).

Another potential route for influencing firm compliance might be the mandatory inclusion of health and safety metrics in annual financial/company reports (for example, refining the inclusion of these metrics within the Global Reporting Index). As Boardman and Lyon (2006) note, numerous reputational factors may influence a firm's engagement with best practice, including: the need to protect/enhance company image/reputation and avoid adverse publicity; and the opportunity to gain a competitive advantage by differentiating themselves within their sector through good health and safety performance or, at least, avoiding being the laggard.

Resource constraints (both finance and personnel attrition) appeared to impinge upon initiatives. Lack of adequate funding is probably the most frequently cited reason for programme failure, particularly by programme operators and client groups. However, this argument must be treated with caution: inadequate funding is a questionable explanation because it is empirically irrefutable – no matter what the level of funding, it can always be argued that it was not sufficient for programme success (Wolman 1981). Nevertheless, resource issues heavily influenced the HSE, and potentially changed the nature of its involvement in the policy process.

## The way forward and the optimal occupational health and safety landscape

Through a series of stakeholder workshops and focus groups, ten key facets of the optimal health and safety landscape were identified, shown in Table 2.3.

Stakeholders broadly agreed that the present regulatory framework is robust and underpins other elements within the landscape. The regulatory framework needs to be in place and implemented in practice through effective inspection and enforcement. The group acknowledged that this may be easier said than done, in light of continued resource cuts to regulators, although more targeted use of intelligence could alleviate this. While the present system was fit for purpose, it was not perfect. It could be more proactive in responding to changes within industry subsequent to the creation of the original regulatory framework (for example, the influx of more SMEs and renationalized industries). With flexibility, though, comes the issue of balance. The system must be flexible enough to cater for the needs of duty-holders, while retaining fairness and consistency. In a related observation, stakeholders thought that the system should be capable of moving more rapidly. New technologies (e.g. Twitter and Facebook) are changing the way people are consuming information.

*Table 2.3* Key facets of the optimal OSH landscape

| |
|---|
| Evidence-based, proportionate and strong regulations which are enforced. |
| Strong, adequately resourced, independent, transparent and competent regulators. |
| The need to elicit a wider range of stakeholder views (e.g. sector associations, trade associations, trade unions, professional associations, etc.) during consultation. |
| A responsive multi-level (i.e. preventive, proactive, consistent and flexible) policy approach which adapts to changes in the business landscape. |
| Competent, open leadership and empowerment and education of management, which facilitates a responsible, strong, top-down culture. |
| OSH is integrated into business thinking and actively championed by business (i.e. OSH seen as enabler). |
| Active involvement, engagement of workforce (not just consultation – behavioural and intellectual buy-in). |
| Access to competent and verifiable OSH support (e.g. guidance, information and advice). |
| More celebration of OSH successes and the promotion of positive messages to the public. |
| Flexible OSH communication plans which are tailored for different audiences (e.g. policy makers, practitioners, the general public). |

Health and safety professionals could be more proactive in using social media to both digest and break news. This would increase the accountability and transparency of businesses, regulators, and practitioners in this sphere.

Although the regulatory framework was generally applauded, its original tripartite working principles had been diluted. Stakeholders felt the system needed to be more open to the active involvement of diverse parties (e.g. trade associations, trade unions, professional associations) throughout proposals to change regulations, especially during consultation phases.

The importance of leaders and managers was noted as a key element within the optimal occupational health and safety landscape. Leaders came in many guises, both within the political sphere and within business. Stakeholders thought that leaders should internalize and then promote the 'value' of health and safety (in terms of keeping people alive, safe and healthy). Leadership is discussed in more detail in a later chapter. Stakeholders should also be advocates to business for the added value of good health and safety management (for example, in increased productivity or enhanced corporate image/ reputation). Health and safety management should be seen as an integral part of business operations (in the same way as, for example, accountancy) rather than an add-on. However, businesses should also be openly voicing the virtues of regulation and assisting regulators in doing their job. By positioning occupational health and safety as a long-term, high-level enabler of business, good management could be seen as an enabler of the wider global economy. This recalls some of the points made in the Introduction and the previous chapter about the idea of a social licence to operate.

Strong leadership should facilitate rather than substitute for the active involvement of the workforce and the public in general. Personal responsibility and accountability was needed from all stakeholders. Although leaders must instil a strong top-down philosophy of good health and safety management, people are ultimately responsible for their own actions, and should not see this as something that happens to somebody else, or as something that somebody else 'does'.

While there was a widespread consensus about the importance of individual responsibility, stakeholders agreed that this needed to be based on competent assistance and advice. Health and safety practitioners have a duty of care to provide professional advice as and when required. In addition, the regulatory authorities, as a basic requirement of their statutory role, should support duty-holders by providing advice and reassurance when needed.

The occupational health and safety system collectively needed to be more vocal in celebrating its successes (i.e. its impressive safety record in comparison to other developed countries). Some felt this would help to counterbalance the negative press. All interested parties must be proactive in educating and lobbying policymakers, businesses and fellow practitioners. This was not to say that a blanket 'one size fits all' positive message should be rolled out to everyone. Messages had to be tailored according to the recipients' varying needs. Different framings are required, for example, at operational and policy levels. The drivers for actions related to health might be quite different from the drivers for actions related to safety.

## Implications of the changing occupational health and safety landscape for stakeholders and practitioners

In the changing landscape, promoting a holistic view, working in partnership and education are vital to the legitimacy of health and safety and the promotion of good practice. In these areas, stakeholders and practitioners can take concrete actions towards achieving a gold standard in occupational health and safety policy.

### Rebuilding legitimacy

As this research progressed, its stakeholders increasingly agreed that the professions and organizations engaged with health and safety needed to do more to celebrate their successes. This echoes the findings reported in the previous chapter. It is also consistent with a recent IOSH salary and attitudes survey, whose respondents thought that industry should celebrate its consistently low fatality rates, and focus on positive health and safety messages (IOSH 2012). Social media and social marketing could be used more intelligently to convey the health and safety system's flexible and dynamic approach. As Lavack et al. (2008) note, the use of social marketing within health and safety is in its nascent stages. There is relatively little within the health and safety literature about this medium's potential role in reducing workplace injury and whether specific initiatives have successfully addressed and reduced occupational injuries. However, the handful of studies (e.g. Spangenberg et al. 2002; Guidotti et al., 2000; Vecchio-Sadus and Griffiths 2004) suggest that the chances of successfully facilitating behavioural change are much higher with sustained and comprehensive social marketing programmes. Health and safety stakeholders and practitioners can play a leading role in rebuilding legitimacy by widely publicizing success stories and launching a dynamic social media strategy.

### Promoting a holistic view of occupational health and safety

A number of new and emerging risks within the UK working landscape are presenting challenges for the regulatory community. These, for example, relate to changes to the industrial landscape, business management trends, the labour force, and human resource management techniques. These new and emerging risks include psychosocial risks, which have high associated costs and are prominent in the modern health and safety landscape. However, the stakeholders repeatedly highlighted a continuing preoccupation with occupational safety to the detriment of occupational health, both by policymakers and practitioners. They criticized the impact of this narrow focus on policymaking and the use of evidence: a frequently cited example was the crude definition of high- versus low-risk sectors in terms of safety rather than health. They also criticized the knowledge and skills of practitioners in relation to occupational health. There was a need to 'put back the H in health and safety'. The professions and organizations involved in health and safety must expand their coverage of occupational health issues, including investments in training and research, and promote a holistic view of health and safety according to the evidence base.

### Supporting evidence-based policy and practice

The Young review repeatedly confused 'low hazard' with small businesses. This has not helped those trying to interpret and make sense of its recommendations. IOSH noted that the two UK sectors with the highest number of fatalities and serious accidents are construction and agriculture, both of which are dominated by small businesses (IOSH 2010). Within our stakeholder workshops, a critical, but still unanswered, question was how activities within a workplace, sector or industry, could be described as 'low risk' without evidence to support this label. Governments have long had robust independent evidence available to them on this question. Our research shows, however, that they have consistently overreached in their response to the recommendations of reviews. The use of evidence in policymaking is critical for policy success. Health and safety stakeholders should continue to play a leading role as providers and resource partners in supplying this. Health and safety practitioners are also critical to supporting evidence-based approaches. They need to maintain awareness of changes in the health and safety landscape through continuous professional development. Leka et al. (2006) investigated future priorities in training provision for health and safety practitioners. They identified needs for a better understanding of risk, legislation and the multifaceted nature of ill health. Practitioners also needed to improve their skills in making the business case for workplace health, and in 'soft' areas like influencing and leadership. Training and education may be a suitable vehicle for providing such knowledge and skills (Leathley 2013). The next chapter investigates some of the challenges that practitioners currently face in accessing relevant and authoritative knowledge and developing appropriate skills.

### Education in occupational health and safety

A key concern emerging from this research was the scarcity of UK higher education courses that are either specific to occupational health and safety or which contain relevant content. This parallels continuing EU efforts to promote health and safety education. The 2003 Rome Declaration *Mainstreaming Occupational Safety and Health (OSH) into Education and Training*, for example, aims to prepare and sustain people during their life, and engage schools and other professional training institutions in actions providing a safer and healthier workforce (EU-OSHA [no date]). The European Commission's *Community Strategy 2007–2012 for Health and Safety at Work* re-emphasized the importance of integrating occupational health and safety into education (EC 2007). Following these approaches, a 'whole-school approach' was launched which specifies how the aims of the Rome Declaration will be reached (EU-OSHA 2013). Löfstedt (2011b) notes that, although demand is high, there remains a dearth of professional courses within European universities. Stakeholders and training providers should supplement their own training activities by promoting collaborations to mainstream occupational health and safety studies in schools and higher education institutions. It is important to educate both business leaders and practitioners and policymakers. As discussed earlier, social media, lobbying, dissemination of research findings and collaborative working with other stakeholders could all contribute to this goal.

### Working in partnership

Perhaps the most prominent success factor in health and safety policymaking, as identified in this research, was developing partnerships with other key stakeholders that balanced diverse interests to achieve shared objectives. In most areas of life (work-related as well as social) collaboration is becoming increasingly common. The need in society to think and work together on issues of critical concern has increased (Austin 2000). This shifts the emphasis in performance from individual efforts to group work, from independence to community (Leonard and Leonard, 2001). Our analysis has demonstrated that multiple stakeholders could work collaboratively in the development and implementation of some initiatives. Both traditional stakeholders, and the newcomers identified in this research, should continue to work together and strengthen collaborations. This is particularly important within the political context, to achieve maximum impact and sustainability of efforts and to raise the profile of occupational health and safety. Finally, as the role of health and safety practitioners continues to grow in the current deregulatory and economic climate – often with the addition of new responsibilities for quality and environmental impact – there is a particular need to consider how the profession can maintain high ethical standards. This is especially acute in dealing with the growing SME sector, who are looking to consultants to understand their needs and to give intelligible advice proportionate to the risks involved. The profession must resist the temptation to prioritize its own liability risks by 'gold-plating' advice to clients. Two of the later chapters in this book look more closely at the situations of SMEs within the contemporary health and safety landscape.

### Conclusion

This chapter reviewed the historical development of health and safety regulation in the UK and the current framework of legislation and policy. It highlighted factors that have been particularly influential in driving change and the initiatives that have resulted from these. It also identified wider lessons that can be learned for the professions and organizations involved in health and safety. The health and safety landscape is not independent of wider influences, including social, economic and political, which define how the issues are dealt with, and the nature of work itself. Stakeholders have often been slow to respond to changes in both the context and in working practices. The more complex the landscape becomes, in terms of the influences it receives (and their outcomes) and the actors that emerge, the more flexibility is required in the systems that control it, evidenced by the increasingly diverse (less prescriptive and more goal-setting and risk-based) forms of regulation implemented in recent years. More stakeholders than ever before are active in the health and safety landscape, promoting their own approaches to regulation. The question then becomes one of how best to balance these competing pressures and policy options in order to achieve desirable outcomes. In this changing health and safety landscape, the legitimacy of health and safety, promoting a holistic view, working in partnership and education are vital to the promotion of good practice. These priority areas need to be considered and acted upon by all key stakeholders, including health and safety practitioners, in order to move closer to the optimal landscape identified in this research.

## Key points

- The present structure of occupational health and safety regulation is closely intertwined with the industrial history of the UK. It is largely designed to address the problems of safety in high-risk manufacturing and extractive sectors that are now much smaller and less central to the economy.
- Recent developments have been driven largely by the development of a single European market where all countries are participating on the basis of shared standards.
- There has been an evolution from hazard-based to risk-based approaches – but organizations have continued to look for specific guidance rather than taking advantage of the flexibility offered.
- Organizations are motivated to comply with best practice by a number of drivers: demonstrating compliance with law; displaying good corporate citizenship; and avoiding the financial costs of health and safety failures.
- Stakeholders in occupational health and safety need to adopt a more proactive approach to managing the perceptions of legitimacy through positive messages about their achievements rather than simply responding to critics.

## 3 The use of knowledge in occupational safety and health: from knowledge creation to employee use

*Joanne O. Crawford, Alice Davis, Guy H. Walker, Hilary Cowie and Peter J. Ritchie*

### Introduction

One concern to emerge from the research discussed in the two previous chapters is the degree to which the professions and organizations involved with health and safety are basing their practice on reliable and valid knowledge. Is their work informed by the best available evidence? Is it really more robust than the much-vaunted 'common sense' of the ordinary citizen or entrepreneur? Should we trust and respect safety professionals as gatekeepers to expertise rather than dismissing them as self-interested peddlers of quack remedies? Our contribution explores how health and safety knowledge in the UK reaches the professionals and organizations that might make use of it. This is a study of the translation of knowledge, of the media used to distribute it and of the ways in which it reaches practice. For the most part, we are dealing with what is known as 'explicit' knowledge: the two projects by colleagues from Loughborough University, described in the following chapters, give more attention to the role of 'tacit' knowledge. We focus on the higher-level questions raised by UK critics of health and safety interventions. Is a body of good quality knowledge available to inform interventions? Can practitioners effectively access knowledge relevant to their concerns in usable forms? Are health and safety practitioners, and the organizations where they work, able to make use of this knowledge?

### Defining knowledge

Before we discuss how knowledge moves around in the field of health and safety, we need to understand a little more about what is meant by the concept of 'knowledge'. Philosophers have been trying to define knowledge for as long as there have been philosophers. A branch of philosophy known as epistemology deals with this topic and is fundamental to all academic disciplines – from physics to sociology. For the purposes of this chapter, however, we shall try to reduce the complexity of the issue by focussing on what it means for the people and organizations involved in moving knowledge around. This will also serve as a background for some of the research described in the next two chapters.

Davenport and Prusak (1998: 5) define knowledge as:

> a fluid mix of framed experience, values, contextual information, and expert insight that provides a framework for evaluating and incorporating new experiences and information. It originates and is applied in the minds of knowers. In organizations, it often becomes embedded not only in documents or repositories but also in organizational routines, process, practices, and norms.

The effective movement of knowledge is more than just passing on pieces of information. Giving workers leaflets or putting up posters at worksites does not prevent all accidents and injuries.

Six characteristics of knowledge are important in understanding health and safety practice:

- knowledge is more than merely data or information;
- knowledge exists in many forms;
- knowledge is credible;
- knowledge is dynamic;
- knowledge must be shared to be useful;
- knowledge is contextual.

### Knowledge is more than merely data or information

Value is added when raw data are processed and transformed into information (Zeleny 2005, 2006; Senapathi 2011). If this information is applied to new contexts, it gains further value from its translation into professional or enterprise-specific knowledge. During the 1960s, for example, women began to appear sporadically among cohorts of people dying from asbestos-related diseases. This is *data*. Further investigations, during the 1970s and 1980s, revealed that many of these women had been employed in factories making gas masks during the late 1930s and World War II. Connecting the data in this way turned it into *information*. Although gas mask manufacture had ceased, this information contributed to a body of *knowledge* that was used to develop increasingly stringent management of asbestos exposures in manufacturing and construction. Eventually, the knowledge was further transformed, into *wisdom*, a theoretical understanding of the basis of the hazard, which enabled it to be applied in contexts quite remote from the original source. One example might be the development of official advice about the safe handling of gas masks in museums and schools. Figure 3.1 shows the continuum from gathering unconnected items of data to their assembly into a coherent whole applicable to novel contexts.

### Knowledge exists in many forms

There is a long history of attempts to define different levels and forms of knowledge. The most important influence on current thinking in organizational

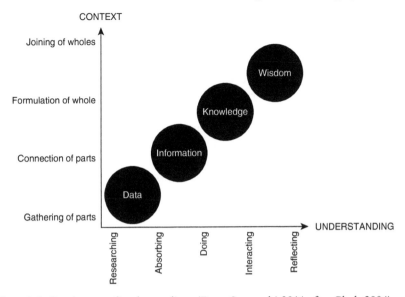

*Figure 3.1* Continuum of understanding. (From Senapathi 2011 after Clark 2004)

*Table 3.1* Five levels at which knowledge resides (Collins 1993)

|  | Knowledge type | Explanation | Example |
|---|---|---|---|
| Tacit | Embrained | Conceptual and cognitive skills | High level OSH knowledge |
|  | Embodied | Action orientated | Safe interactions with environment and people |
|  | Encultured | Shared understandings and norms | Language and safety culture |
|  | Embedded | Routines and guidance | Formal OSH/Health and Safety procedures |
| Explicit | Encoded | Stored knowledge | OSH databases and knowledge repositories |

studies about knowledge management is the work of Nonaka and his col-leagues (Nonaka 1994; Nonaka and Takeuchi 1995; Nonaka et al. 1996). They elaborate on a well-recognized contrast between tacit and explicit knowledge. This may, though, be more of a continuum than a contrast, as suggested by Collins (1993), a leading sociologist of science and innovation (see Table 3.1). Tacit knowledge is revealed through the practical skills required for the per-formance of a task. These skills are often manual, as where a carpenter assesses the challenge of working with a particular piece of wood. However, in some occupations, they can be largely cognitive. An experienced health and safety practitioner, for example, can walk into many different situations and iden-tify potential generic hazards and management strategies without needing context-specific expertise. This is sometimes described as 'professional vision', although it is not necessarily linked to the status of the person who is doing the seeing (Goodwin 1994). Police officers on patrol, for example, learn how to monitor streetscapes, for potential crime, in a quite different way to civilians who are going about their everyday business (Sacks 1972). Nurses trained in back injury prevention learn how to scan a task in a new way to identify potential lifting hazards and appropriate solutions (Policy and Strategic Projects Division 2002). Such knowledge forms part of the local working culture.

Tacit knowledge can be contrasted with the explicit knowledge about hazards and their management stored in forms such as scientific journals, company databases, operating manuals or accident reports. This knowledge is more formal, systematic and easily shared. It is 'out there' rather than in the minds of the users. Past events and observations are assembled into context-free theories expressed in rational forms of language. Because they are now detached from the particularities of time and place, they can be made into objects of reflection and debate. Logical structures and order can be imposed upon these data. Knowledge, or wisdom, becomes public and portable.

### Knowledge is credible

Knowledge also has to be credible. This means that it must have been tested or tried out in some way that distinguishes it from opinion, belief or speculation (Murray and Peyrefitte 2007). However, credibility is never fully independent of the context in which the knowledge is created and received. Where it points towards actions that are costly – in economic, political or reputational terms – the authority of knowledge may be contested for long periods, as we have seen with the link between smoking and lung cancer or the debates over climate change. In the context of asbestos, for example, there was a lengthy argument between clinicians and epidemiologists about whether there was a causal relationship between exposure and lung cancer, independent of the prevalence of smoking among asbestos workers. On the basis of their, tacit, clinical judgement, many clinicians held that asbestos did not cause lung cancer: exposed workers often smoked, so this must be a spurious association. However, epidemiologists, using explicit, statistical, methods determined, at a population level, that non-smokers exposed to asbestos also had a significantly enhanced rate of lung cancer. Both tacit and explicit knowledge had been tested according to the methods considered appropriate by their users and found to be credible. In this instance, the contest was eventually settled by the judgements of courts on liability for the health consequences of asbestos exposure, which preferred the explicit knowledge of the epidemiologists to the tacit knowledge of the clinicians.

### Knowledge is dynamic

Research and publication, in their broadest sense, gradually transform tacit knowledge into explicit knowledge. The tacit knowledge of clinicians – only men get asbestos diseases – is first challenged by the appearance of women patients and then revised by research. Such transformations involve processes like replication, peer review, presentation to fellow specialists and eventually to other professionals. If successful, these processes lead to an agreed position that can be translated into guidance documents for use by people with less specialized knowledge. The more explicit knowledge is, the more easily it can be shared and moved around within an organization or between organizations (Kang et al. 2010). Studies of the aviation industry, for example, have created explicit knowledge about high-reliability organizations that has been taken into healthcare in an effort to improve patient safety. Rather than externalizing the tacit knowledge of their own employees, hospitals have imported explicit knowledge and promoted its internalization.

Explicit knowledge is not necessarily stable: it is knowledge that works in a particular context with particular technology. The safety design of motor cars, for example, is a mature field with well-established and widely-shared explicit knowledge about how to manage major hazards. Manufacturers now have to adapt to the implications of self-driving vehicles, for both the occupants of those vehicles and those in traditional cars who will interact with them. Explicit knowledge must move between companies that specialize in software, GPS systems and cybersecurity, and companies with the traditional skills involved in powertrain design, bodywork and passive safety measures.

### Knowledge must be shared to be useful

Knowledge is not a useful resource in isolation. If it is to create value, it must be shared within, and sometimes between, organizations (Kang et al. 2010). Part of the management challenge for an organization is to decide how much to invest in developing or acquiring health and safety knowledge. Should the organization be doing its own research or accessing research produced by other people? Does this involve paying to license a product innovation or course fees for some, or all, staff to undertake further training? Can the organization exchange its own knowledge for knowledge produced by a collaborator, or even a competitor, which is quite common in some high-technology industries? The answers will be different at different scales – a large organization can afford to do its own research and development more easily than a small one. Some organizations, of course, may specialize in doing research and development on behalf of other organizations that do not have the resources or specialist skills to act on their own.

### Knowledge is contextual

Knowledge is a dynamic mix of experience, contextual information, values and expertise. As we have seen, knowledge does not stand alone. Its value is created in the movement from the context in which it is gained to the context where it is applied. This process is shaped by a variety of factors in the social, cultural, economic and political environment.

As these six key themes highlight, knowledge is more than just bits of information. It exists in many forms, is stored in different places and only brings value when it moves within or between organizations. Although both tacit and explicit knowledge are important for organizations, the latter is often thought to be more authoritative. While the research reported in the next two chapters questions this assumption, our brief was to look at the relationship between the health and safety profession's practice and the explicit knowledge available to it. Was the profession making the best use of what some critics thought was the 'best evidence' – or was it just making things up as it went along, to the annoyance of those trying to do 'real work'?

## Moving knowledge between organizations

When scientific knowledge moves out of the laboratory in the form of innovations, these still have to be taken up in society before anyone can benefit from them. A recent example might be the application of genetic modification technology to important food crops. This innovation is based on outstanding scientific achievements but its impact has been patchy because the social process by which the products were introduced to consumers was poorly managed.

The processes by which knowledge moves between people or organizations were first studied by social scientists during the 1960s. This research was driven by two main factors: a perception that the rate of innovation had become much faster since World War II; and a growing expectation that scientific work should be useful to society. Were we inventing things more quickly than we could make effective use of them? How could we alter a situation where great science was not leading to great products or changes in practice? Concern for 'technology transfer' was given greater urgency during the 1970s by the rise of global competition and the pressures to develop national innovation systems that would make the most productive use of public and private investments in research and development. Early studies focussed on the movement of material products or technologies. How did farmers decide to plant new crops or doctors to prescribe new drugs? However, this came to be regarded as a special case of the more general question of how knowledge moved. We might, for example, look at the way in which Deming's (1982) ideas about quality assurance were taken up around the globe using the same approaches that we would use to look at how cancer specialists chose to prescribe a new medication.

While early researchers (e.g. Zander and Kogut 1995) tended to talk about 'knowledge transfer', terms like 'translation' and 'transformation' came to be preferred (Graham et al. 2006). 'Transfer' implied that the movement was a simple linear process between organization A and organization B. 'Translation' recognized that the knowledge was likely to have to be modified in the course of this journey if organization B were to be able to make use of it. 'Transformation' acknowledged that the process went further than simply restating the knowledge in terms comprehensible to organization B. It might involve some qualitative change so that explicit scientific knowledge became tacit or was embedded in material objects (Carlile and Rebentisch 2003).

Handwashing practices, for example, may be based on chemical or microbiological knowledge but need to be adopted as routine and unthinking behaviour by relatively low-level workers. The knowledge is transformed from a scientific understanding to a simple habit. On the other hand, there may be circumstances where this is not sufficient. Our colleagues from Loughborough have documented how community health professionals improvise hand-cleaning procedures to minimize the risks of cross-infection to themselves and their clients. These workers use their knowledge of microbiology to assess the risks and choose ways of managing them that suit the unpredictable and changing contexts of different clients' homes (Pink et al 2014b). Alternatively, consider the ways in which plugs on domestic appliances are now moulded on at the factory so that consumers no longer need to know how to wire them correctly. Safety knowledge has been embedded into the design of the object instead of being explicitly conveyed to the user.

Several large studies have examined knowledge translation within medicine. However, little is currently known about how these processes might operate in the health and safety field. Work by van Dijk et al (2010) gives an indication of the importance of a knowledge infrastructure for health and safety professionals but this study does not extend to the non-professionals implementing health and safety initiatives in the workplace, and the knowledge translation tools that they might need. Roy et al. (2003) summarize some issues that have been identified in health and safety knowledge translation. One is that health and safety researchers and practitioners are separated by the way in which they consider knowledge. Knowledge from research is difficult to move into existing organizations through peer-reviewed publications, which, understandably, target an audience of peers. Researchers produce explicit knowledge, focussing on specific questions, taking a neutral stance, and dealing with few variables. Practitioners, however, are often concerned with solving current and immediate problems, comparing explicit knowledge with their tacit experience, and dealing with multiple variables in large and complex systems. The adoption of new knowledge occurs within the pre-existing constraints of an organization, creating a potential for it to clash with existing ways of working. Sometimes, research that emerges from one industrial context will be ignored in others unless it is repurposed to make its relevance clear. It must be translated into a more practical form to be usable in applied contexts.

Reflecting this thinking, Canada, for example, has developed 'Knowledge Networks', to support health and safety research and practice (Canadian Institute for Health Information, 2006). Based on survey data from 217 researchers participating in these networks, Laroche and Amara (2011) found that the movement of knowledge from scientific research to practice was more positively associated with the publication of research reports and personal connections with end users than with the number of peer-reviewed articles. Although these articles might be important for other reasons, Laroche and

Amara concluded that the most important facilitators of knowledge movement were its adaptation to the capacity of end users to absorb it, the level of resources dedicated to dissemination and the support for networks linking researchers and users. End users, in this case health and safety professionals, need to be engaged throughout the knowledge development process if this is to have a positive outcome.

The field of occupational health and safety has already absorbed the knowledge that *communities of practice* are important (Lave and Wenger 1991). These link people with common skills or interests across internal or external organizational boundaries. They include social networks (personal contacts, professional networks and social media), which are described as scale free because they can increase in size dramatically, yet remain accessible by all members of the network. The same networks have been found to facilitate the movement of knowledge (Tang et al. 2006; Lin and Li 2010).

Much research has focussed on the people or organizations responsible for supporting these processes of transfer, translation or transformation. Two roles are important: *knowledge broker* and *boundary spanner*. These terms are often used interchangeably but there are some useful distinctions. Knowledge brokers are individuals or groups who occupy niches that specialize in collecting and transforming knowledge. They manage or systematize the knowledge, linking its producers and its users and building the capacity of both to exchange and absorb each other's knowledge (Meyer 2010). This process can also be seen in relation to new legislation where a law may be enacted, but its practical meaning and impact are achieved through the interpretations of intermediaries, who may have interests of their own. Legislation may set out to provide flexibility in use but this progressively disappears as it is translated into operational procedures by people who are aware that they may be sued if these procedures fail. Boundary spanners may be thought of as people or groups who work within organizations to reach out across internal or external boundaries and make connections that join up otherwise distinct areas of knowledge and expertise. They may be the clients or partners of knowledge brokers, or they may take some of this function upon themselves (P. Williams 2002). Health and safety professionals may be found in either role.

## Who are the health and safety knowledge brokers?

In the UK, substantive, explicit, health and safety knowledge originates from a disparate range of sources. By 'substantive knowledge' we are excluding simple lists of intermediaries, like directories of trainers or experts, raw data sets, policy statements and informal communications through online forums. While all of these may be considered to be explicit knowledge in a general sense, the focus of our study was on sources that directly present or translate research evidence for practice. In responding to public criticism of the legitimacy of the profession's use of knowledge, it is particularly important to look at how the most robustly tested forms of knowledge are taken up by practitioners.

We identified 303 sources of knowledge brokerage online – many of these also exist in document form, either in print or as downloadable text. While the decision to focus on online sources was partly one of convenience, this also provides the most comprehensive sample: documents tend to be available online but online materials are not necessarily available in print. We collated and summarized the topics that were covered and the formats in which the knowledge was made available. Many of the sources were government departments, agencies and authorities. Trade associations were also important, although their websites varied in credibility, in the sense discussed earlier. Professional societies and groups provide information to their members and, in some cases, to the public. The need for professional accreditation and continuing professional development by health and safety professionals is one route for knowledge translation within occupational health and safety. Some trade unions provide specific health and safety information and guidance to their members, as do some employer organizations. A number of charitable or 'not-for-profit' organizations were identified in specific sectors. There were a few voluntary organizations and a number of private companies: the latter usually required a membership subscription. Trade magazines provided a further source. Other categories included some organizations that were directly linked to health and safety issues, while others were less central, although providing some relevant information, like the *Equality and Human Rights Commission.*

The health and safety knowledge provided by these brokers included guidance, legislation, primary research reports, systematic evidence reviews, training materials and presentations, delivered through such media as print publications, web publications, DVD/video, leaflets and posters. The topics covered can be categorized into: 'hazards', such as electrical safety and working at height; 'health outcomes', including musculoskeletal disorders; and finally, 'general topics', including health promotion and construction (Figure 3.2). Research into safety and/or health was funded, and new knowledge generated, by government departments, universities and independent research organizations. However, much of this work was targeted on very specific safety questions or health effects and focussed on particular industrial contexts, as Roy et al. (2003) also found. Focus groups, which we conducted with experts and industry representatives, identified particular gaps in the online knowledge base in relation to health outcomes.

## How do knowledge brokers assess the quality and value of knowledge and information provided?

A sample of 16 knowledge brokers were contacted to ask about their procedures for assessing the quality of the knowledge and information that they provided. Most, but not all, were able to describe an internal review process that preceded publication. Our focus groups suggested that information from some 'official' sources, such as HSE, IOSH, TUC and academic journals, is automatically trusted to be accurate and of high quality. Information from

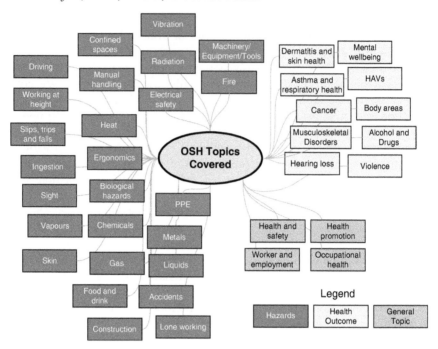

*Figure 3.2* Map of the topics covered by the information providers

other sources would be checked by the individual brokers who wrote the content for an organization's website.

Organizations generally had some data on how many times a website was accessed and how long individuals might spend on a particular theme. However, this type of information yields relatively little information on end users: organizations did not, for example, log IP addresses to map the location of users. Some websites required registration to download or access specific areas but providers made limited use of this information to guide them on the presentation of content or to identify 'hot topics'. There is, of course, something of a trade-off here: asking users for more feedback can be a disincentive to their engagement with the site. While there can be little control over the ways in which knowledge is used once it has been made publicly available, the knowledge brokers surveyed did identify a number of possible follow-up methods. These include: feedback after material had been used in training and workshops; identification of how processes and procedures have changed due to new knowledge provision; and the use of other forms of engagement such as helplines.

It is apparent that health and safety knowledge brokers are much more focussed on the delivery of knowledge than on its reception. While the knowledge may be quality-assured, there has been less consideration of its value to the intended audience, the choice of media and the degree of uptake.

Communication tends to be one-way, with limited use of the feedback that is available from site user data, let alone more deliberate attempts to ascertain what the audience needs and how best to encourage them to incorporate new knowledge into their own practice.

## Who are the health and safety boundary spanners?
## Which brokers do they use?

In an attempt to fill this gap, we undertook a survey to identify the sources that were actually being used by people with a boundary spanning responsibility for occupational health and safety. There were 386 eligible respondents, comprising 84 health and safety consultants and 302 employees (Figure 3.3). 'Employees' were defined as individuals who are employed as safety or health practitioners or individuals tasked with health and safety matters within their organizations. 'Consultants' may, at times, be either knowledge brokers or boundary spanners. In this survey, most of the consultants were operating on a relatively small scale and further translating materials from brokers for client-specific use. It seemed more appropriate to treat them alongside employees fulfilling the same function, although some differences do emerge as a result of their different structural positions.

Amongst the employee respondents, 65 per cent were professional practitioners in their own workplace. Of the consultants, 80 per cent had 11 or more years' experience in the field, compared to 62 per cent of employees. Of those who completed the survey, 91 per cent of consultants and 84 per cent of employees were members of at least one professional organization, with the most frequently reported being IOSH, *the British Occupational Hygiene Society*

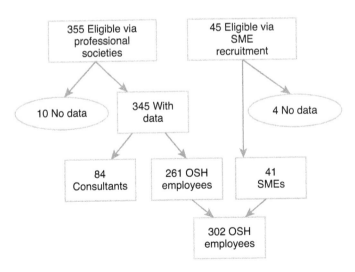

*Figure 3.3* Distribution of eligible survey respondents

(BOHS) and the *International Institute of Risk and Safety Management*. Most employees and consultants had one or two health and safety qualifications, most commonly a NEBOSH National General Certificate (n = 145) and/or a NEBOSH diploma (n = 144).

The survey found that government knowledge brokers (e.g. NHS, HSE) were the most widely used source of health and safety knowledge, by both consultants and employees, when obtaining information on hazards or safety issues. The reasons mentioned included the level of trust in the source, ease of access, free access and clarity of presentation. This indicates that respondents recognized the importance of seeking authoritative information so that they could appropriately advise their organizations. Respondents identified a total of 128 individual websites, of which the HSE and IOSH websites were most often reported to be used. The reasons given for these preferences were that these sites are seen as relevant, informative, up-to-date, trustworthy, authoritative, professional and unbiased. In general, Internet searches for health and safety knowledge were preferred as they are seen to be quick, relevant and offering a choice of information to read and follow. This validated our own decision to focus on knowledge brokers with an online presence.

## The impact of knowledge

In order to understand how explicit knowledge gets taken up by organizations, we adapted an approach from diffusion of innovations theory (Rogers 1983). Although, as we noted earlier, technological innovation is not exactly the same as knowledge, the relationships described by Rogers are derived from comparable research studies and rest on similar principles and processes. This approach directs us to examine four elements: the content of the knowledge; the communication channels (media) used; the time span for translation; and the system through which knowledge is communicated (see Table 3.2).

Figure 3.4 presents a simplified model of how knowledge (content) can flow from its source in a research base through the medium of a professional magazine (knowledge broker) to health and safety specialists (boundary spanners) and then into an organization. In practice, of course, the information might flow through other media associated with a community of practice – conference presentations, local meetings of professional associations, even conversations by a coffee machine or on a golf course. The people compiling the research, or evidence, base will also have other interactions, as this research programme itself shows. Some of the information will come from researchers in medicine, science or engineering. Other information will come from studies of how people at different levels do their jobs and how their work is located within organizations. There will also be research evidence on the communication processes themselves. This involves some reference to the 'richness' or intrinsic qualities of the media, as well as to the culture and structure of the organization through which the knowledge must pass.

*Table 3.2* Senapathi's (2011) critical elements for successful knowledge transfer

| Term | Definition | Issues in effective dissemination |
|------|-----------|-----------------------------------|
| Source | The dissemination source, that is, the agency, organization or individual responsible for creating the new knowledge | Perceived competence: Credibility of experience and motivation Sensitivity to user concerns Relationship to other sources trusted by users |
| Content | The message that is disseminated (tacit/explicit) | Credibility of research and development methodology: Credibility of outcomes Cost effectiveness Relationship between outcomes and existing knowledge |
| Medium | The ways in which the knowledge is described and transmitted | Accessibility and ease of use, user friendliness: Flexibility Reliability Cost-effectiveness |
| User | User or intended user | Perceived relevance to own needs: User's readiness to change Information sources being trusted Dissemination media preferred |

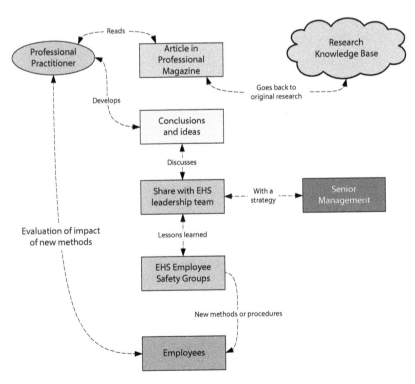

*Figure 3.4* Theoretical model of the process of knowledge transfer

The idea of 'richness' provides a way to recognize the effects of different kinds of media. Rich media tend to be personal in nature, offering multiple cues and immediate feedback. Face-to-face interaction is typical. A good deal of health and safety knowledge, however, moves through lean media such as documents, which simply lay out explicit rules, policies and procedures. These media are impersonal and offer few opportunities for interaction or feedback. Neither type of media is necessarily and invariably better than the other (Murray and Peyrefitte 2007). The choice depends on the context within which communication is taking place and the objectives of the communicator, as shown in Figure 3.5.

Unequivocal or unambiguous messages can be adequately transmitted through lean media by means of documents or procedures. In some contexts, strict adherence to protocols is essential for safety and messages are simply about sticking rigidly to the protocol. However, in other contexts, a degree of judgement is called for and information is inevitably more equivocal or uncertain. Rich media, such as face-to-face contact or training scenarios, may be a more effective way of moving knowledge in these circumstances.

Communication, of course, is not just a matter of sending messages – they also have to be received. The addition of organizational culture and structure as variables allows us to deal with the *absorptive capacity* of individuals or organizations (the ability to recognize the value of new knowledge, the skills to assimilate it and the motivation to do so). Higher levels of absorptive capacity facilitate the translation of knowledge into new social and cultural contexts.

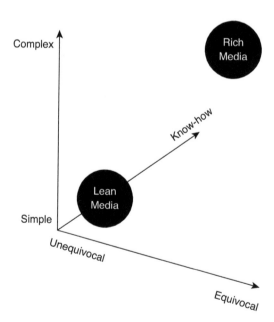

*Figure 3.5* Model of contingency factors and their relationship to lean and rich media

Absorptive capacity is a property of both individuals and organizations: individuals must be open to receiving new knowledge, and capable of using it, but can only do so within organizations whose structures and cultures make this knowledge available and support its incorporation into practice.

The role of organizational culture, as a dimension of the context through which knowledge moves, is particularly critical. One element of this is the notion of 'safety culture', which is already widely accepted within the professions and organizations involved with health and safety. Cooper (2000) notes that safety culture has three interrelated aspects: psychological, behavioural and situational. The psychological aspects refer to how people within organizations feel about safety and safety management systems. The behavioural aspects are concerned with what people do within the organization and the situational aspects refer to the policies, procedures and measures the organization has in place. The movement of knowledge is shaped in similar ways by individual receptiveness, by behaviour, and by organizational policies.

The organizational structure through which knowledge is being moved also has an impact. Within centralized and hierarchical organizations, for example, knowledge implementation can be very efficient: one individual can absorb knowledge and direct the activities of other individuals. However, peer-to-peer networks, where decision-making is dispersed, seem to be a better structure for knowledge generation and sharing. There is more feedback and collaboration in problem-solving, engaging people and securing active commitment to using knowledge rather than being passively directed. This may be particularly useful where work environments are unpredictable. The Loughborough research, for example, notes the challenge to logistics workers making home deliveries of bulky items. The delivery teams must be able to improvise from their tacit knowledge about safe handling, while warehouse workers may be able to rely on explicit protocols to handle standard items in a controlled environment.

Five core questions ultimately influence organizational or individual decisions about whether or not to adopt new or translated knowledge:

- What does the translated knowledge contribute? (Relative advantage.)
- How easy will it be to access and assimilate the new knowledge? (Compatibility.)
- Is the effort involved in using the new knowledge worth it? (Difficulty.)
- Can end-users experiment with the knowledge? (Trialability.)
- Is the new knowledge visible to others? (Observability.)

## What do boundary spanners look for?

When looking for new information on health and safety, the format of information most often sought, by both employees and consultants, was guidance documents (see Table 3.3).

Guidance documents and legislation are used most often by employees and consultants in keeping up to date with regulations and best practice (Table 3.4).

*Table 3.3* Formats used to find information about a new hazard or health outcome

| Formats | Number of OSH employees | Number of OSH consultants |
| --- | --- | --- |
| Guidance | 249 | 71 |
| Legislation | 224 | 67 |
| Professional magazines | 185 | 57 |

*Table 3.4* How respondents keep up to date with OSH information and knowledge

| How they keep up to date with OSH | Number of OSH employees | Number of OSH consultants |
| --- | --- | --- |
| Guidance | 224 | 67 |
| Legislation | 223 | 65 |
| Internet searches | 175 | 44 |
| Magazines | 166 | 55 |

Respondents noted that the usability of health and safety knowledge can be affected by various design issues, including font type and size, language, content, terminology and communications media used. At an individual level, the sources of information reflect an individual's career progression. With growing experience and wider professional networks, more experienced employees and consultants become more embedded in a community of practice. They make increasing use of personal contacts within an industry to discuss issues, rather than depending on formal sources of information. This reflects the challenge of translating even practical guidance documents into everyday ways of working. It may be easier to learn from someone who has already done this and can describe their personal solutions. Of course, this may as easily become a way to spread bad practice as good, following over-prescriptive or risk-averse interpretations of guidance as well as more flexible ones! It raises questions about how well communities of practice actually work in the field of occupational health and safety.

## How do boundary spanners share knowledge within an organization?

Communicating knowledge to others is a key part of the role of health and safety employees and consultants. The survey, then, asked how relevant knowledge was selected and shared with others. For all topics, health and safety employees most frequently chose, as means of dissemination, meetings, emails, training courses and toolbox talks, while consultants were slightly more likely to use the Internet/intranet and less likely to use toolbox talks (see Table 3.5). This may reflect the different structural positions of those employed *in* the business and those employed *by* the business. Toolbox talks may be more useful when delivered by someone known to staff members, than by an external consultant.

*Table 3.5* Methods used by OSH employees and consultants to communicate OSH information for different issues. Each cell shows percentage of those responding to each question

| | Hazard | | | | | | | | | |
| | Physical | | Chemical | | Biological | | Safety | | Health | |
| Method | E | C | E | C | E | C | E | C | E | C |
|---|---|---|---|---|---|---|---|---|---|---|
| Meetings | 77% | 76% | 72% | 77% | 69% | 71% | 78% | 72% | 81% | 71% |
| Email | 71% | 85% | 68% | 85% | 68% | 84% | 74% | 86% | 74% | 84% |
| Courses | 69% | 61% | 73% | 67% | 68% | 63% | 74% | 68% | 63% | 61% |
| Toolbox | 65% | 44% | 65% | 43% | 58% | 34% | 63% | 49% | 49% | 35% |
| Internet | 54% | 50% | 57% | 50% | 59% | 47% | 58% | 54% | 59% | 53% |
| Newsletters | 36% | 35% | 38% | 35% | 40% | 42% | 42% | 39% | 40% | 39% |

E = OSH employees, C = Consultants

*Table 3.6* Methods used by OSH employees and consultants to communicate OSH information to different levels of employees. Each cell shows percentage of those responding to each question

| | Level of employee | | | | | | | |
| | Senior management | | Middle management | | Employee | | New start | |
| Method | E | C | E | C | E | C | E | C |
|---|---|---|---|---|---|---|---|---|
| Meetings | 87% | 83% | 88% | 80% | 62% | 52% | 50% | 52% |
| Email | 83% | 85% | 85% | 83% | 58% | 50% | 36% | 34% |
| Courses | 44% | 32% | 65% | 52% | 78% | 73% | 86% | 86% |
| Internet | 43% | 37% | 51% | 37% | 50% | 32% | 37% | 36% |
| Newsletters | 27% | 24% | 34% | 22% | 42% | 33% | 31% | 30% |
| Toolbox | 14% | 7% | 27% | 15% | 66% | 52% | 50% | 45% |

E = OSH employees, C = Consultants

A similar pattern was identified in the methods used to communicate information to different audiences (Table 3.6). Senior managers were more likely to be given information through meetings and email. However, when communicating with employees and new starts, training courses and toolbox talks are used most often by health and safety employees, and meetings and training courses by consultants. Again, this may relate to their different structural relationship to the company: training may be more formalized if an external consultant is brought in.

When asked about their own preferred ways of communicating health and safety information to others, employees preferred courses, followed by meetings and toolbox talks, whereas the consultants preferred meetings, courses and email. Again, this probably reflects their different structural relationships with a company. When asked about the most effective ways of communicating, employees listed toolbox talks, face-to-face communication, training courses, health and safety committees and one-to-one coaching for professionals. Consultants provided a similar list but also considered email communication as effective. The majority of the most frequently used and preferred methods of communication involve rich media – direct contact with others, through training or other types of face-to-face interaction. This may reflect the complexity of the knowledge being translated. It may not be appropriate to communicate this through lean routes such as the Internet/intranet or email.

Respondents were also asked whether they evaluated knowledge translation and its implementation through workplace changes (Figure 3.6). This was most frequently achieved, according to both employees and consultants, by talking to employees and completing safety inspections. Differences in approach were identified in relation to company size: smaller companies spent more time talking to employees, using safety inspections, with risk assessments (and re-assessments) and injury monitoring being used more by larger businesses. This reflects the more formal processes within larger businesses, especially those dealing with higher levels of hazards/risks. These organizations are more likely to have systems in place to monitor and record key data, including sickness absence reporting, days since last accident and maintenance scheduling.

Fewer SMEs, specifically those companies with ten or fewer employees, seemed to be using these methods. However, the reported frequency of talking to employees is high across all company sizes, underlining the importance of face-to-face contact in both implementing and evaluating the impact of change.

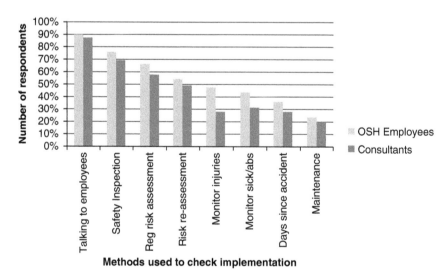

*Figure 3.6* Methods used to check implementation of information. Figure shows percentage of respondents reporting use of each format

The survey also explored barriers to effective communication. Many respondents (n = 85) reported that there were no barriers: this was the most frequent response, showing that they think they have found effective ways of translating health and safety knowledge. Respondents did however identify 48 different barriers to knowledge translation. Of these, the most common were time constraints (n = 28), literacy and language hurdles (n = 18), the culture of the workforce (n = 17), lack of management support (n = 17), not getting people together at the same time (n = 15), lack of interest (n = 13) and geographical spread of the workforce (n = 11).

Issues of literacy and language suggest that those delivering knowledge and information need to understand their audience, if they are to ensure that materials are at an appropriate level to be effective. Those delivering knowledge must understand not only its content but also how it can best be transformed to meet the needs and receptiveness of the target audience. This may seem like an extra cost to a business, but it ensures that resources are not wasted on ineffective communications.

We also asked questions that related to Collins's (1993) five levels at which knowledge resides (see Table 3.1). We found that smaller companies made little use of high-level, embrained, tacit knowledge, which makes cognitive connections to everyday work or process design almost without conscious thought. Nor do they make use of untransformed knowledge such as scientific journal articles. Those who have professional memberships are more likely to refer to legislation, magazines, journal articles and courses than those who do not. This seems to be less a question of access to research material published behind paywalls in academic journals than of absorptive capacity. Smaller companies do not have the resources to invest in accessing and translating material of this kind and rely on knowledge brokers and boundary spanning consultants to do it for them.

When it came to their own preferences for receiving information, respondents also endorsed face-to-face methods. While they consulted websites and documents, they thought that talks and training sessions were more effective because they were more interactive and targeted to the needs of a specific audience. This is consistent with the conclusions of the review by Hasle and Limborg (2006) and the research by Roy et al. (2003) and Laroche and Amara (2011), which point to the perception that face-to-face communication is more trustworthy and to the different understandings of what constitutes knowledge between researchers and practitioners. It is notable that, while boundary spanners have to engage with lean media, they are trying to pass on the information in richer forms. The value attached to face-to-face interaction is also reflected in the survey's findings about the importance of direct engagement with other individuals and with social networks to obtain new information through communities of practice. In this way a fluid mix of experience can be assembled and drawn upon to acquire health and safety knowledge for a company (see also Davenport and Prusak 1998; Yakhlef 2007). Safety professionals join the input from these personal networks with the specific content of the company to establish knowledge (Senapathi 2011). This does, of course, place a considerable premium on the networking skills of both employees and consultants and on judgements about what knowledge must be protected for commercial reasons and what can be shared in a collective interest.

## How is health and safety knowledge translated in industry?

Surveys are useful in collecting information about what people say they think or do. However, there remains an uncertain relationship between answers to a questionnaire and actions when people actually have to deal with a problem. To address this, we carried out 12 case studies in collaboration with a range of organizations. Each organization was in the process of undertaking a health and/or safety intervention. The case studies were based on a mix of interview and survey data. Although this is not quite as robust as data from direct observation, which features prominently in the next two chapters, it gets much closer to the actual decisions being made, and their rationales, than can a survey. We could ask, 'why did you do that specific thing when this specific thing happened?' rather than asking, 'what do you generally do when things like this happen?' (Further details can be found in Crawford et al., in press 2016). A summary of the case studies is presented in Table 3.7.

## The skills needed by the boundary spanner

The boundary spanners in these case studies were health and safety professionals who worked in specific roles, including health and safety managers and safety officers. These individuals had a high level of health and safety expertise, which included knowing where to obtain new information or who to speak to about possible new hazards.

Early-career safety professionals make more use of textbooks or course materials, while non-professionals seem to use less formal media as a source of information. Smaller companies rely more for safety information on trade media, suppliers or publicity campaigns from sources such as the *Health and Safety Executive*. The education of health professionals seems to give more emphasis to the skills of finding and evaluating explicit knowledge before using it in practice. These skills are still developing within the routes into the health and safety profession from safety, ergonomics and occupational hygiene. This may point to a particular problem for smaller companies, with a limited absorptive capacity, in critically evaluating explicit knowledge for its relevance to their business rather than simply applying it in a cookbook fashion. It may account for some of the contradictions between demands for simple, practical rules to follow and complaints about over-prescriptive and irrelevant interventions.

The boundary spanners within the case studies used their skills and previous knowledge to plan and action policies and strategies to guide the intervention in the company. As part of this translation of knowledge, and the general safety

*Table 3.7* Case studies included in the research

| Company type | Type of case study | Intervention | Size of company |
|---|---|---|---|
| School within a university | Retrospective | Portable electrical equipment safety intervention | Large |
| Catering industry supplier | Retrospective | Introduction of H&S committee | Large |
| Roofing company | Retrospective | Refresher face fit training | Small |
| Skip manufacturer | Retrospective | Re-emphasis of hearing protection through information and new types of hearing protection | Small |
| Engineering and construction | Retrospective | Introduction of a new induction process | Large |
| Construction, engineering and development | Prospective | Introduction of a new type of cable locator | Large |
| Facilities management | Retrospective | Introduction of a health surveillance matrix | Large |
| An aerospace and defence company | Prospective | The impact of change in either/or policies and procedures in relation to working at height | Large |
| Housing association | Retrospective | Introduction of an office safety network (OSN) | Large |
| Fire safety group | Retrospective | Introduction of health and safety policies and risk assessments | Micro |
| International retail company | Prospective | Introduction of a new online health and safety induction | Large |
| Banking and financial services company | Prospective | Health and safety documentation and intranet content redesign as part of the 'Health and safety remediation programme' | Large |

culture at the company, the boundary spanner aimed to get the knowledge accepted as a shared understanding by the target audience.

## How to transform/translate knowledge effectively

From the case studies we found that organizations used both virtual and face-to-face processes to move knowledge to end users. Smaller organizations used the richer media of face-to-face interaction and other informal routes for knowledge translation. This was due to the close connections and easy contacts with employees that are often available in this context: virtual processes were simply not required. Larger companies, especially when spread across different geographic locations, primarily used virtual methods (such as emails and presentations), which tend to be leaner in character.

## What are the success criteria for knowledge translation in occupational health and safety?

The success of knowledge translation depends upon various factors. All the case study organizations had developed a plan before implementing the intervention, although some of these were more informal than others. However, those implementing the intervention had rather different levels of discussion about how they were going to take the work forward.

The boundary spanner's experience can affect the outcome of the process. Experience refers both to their understanding of the relevant knowledge and their ability effectively to change and adapt the format and content of its communication. The effectiveness of knowledge translation relies heavily on the target audience's understanding of the information being disseminated to them. Health and safety professionals must understand the need to translate knowledge into an accessible language, context and reading level, and have, or are able to acquire, the skills to do so. This applies both to the technical level of the information, in relation to what the employees already know about the topic, and to whether the information is meant to be understood as stand-alone new information or as incrementally additional information to extend the employees' existing health and safety knowledge.

The health and safety professional's rapport with the target audience, and the visibility of health and safety as a workplace issue, influence the effectiveness of communication and the quality of its reception. Employee involvement can aid knowledge translation: our case studies used a variety of methods such as workers trialling new tools and equipment (compatibility); piloting an intervention (trialability); taking part in a working group or committee (difficulty); and being workplace safety champions (relative advantage). These activities increase the visibility (observability) of the translation, inviting the involvement of a broader range of employees and helping to establish health and safety as a collective objective.

## What needs to be considered when disseminating information – audience, format, language?

Knowledge translation must be sensitive to the size of a company and the spread of employees who need to acquire the knowledge. In the case studies, larger companies used already existing information routes and structures to convey information provided by health and safety professionals. In smaller companies, although the professional provided knowledge, the closer connections between employees facilitated a more two-way, and often richer, process.

The dissemination of information also needs to be tailored to the safety culture of the company. In our case studies, health and safety professionals and employees seemed to share a common understanding about the status of safety as a goal and to have similar readiness to adopt new knowledge in pursuit of this. We can see these as markers of a positive safety culture.

Other issues that need to be considered include: the target population; the type of work they do; who will be disseminating the information; and how it will be disseminated. This consideration should be influenced by the nature of the knowledge to be translated and the resources available: the translation of complex material will be less successful if sent through the lean medium of an email. These processes require health and safety specialists to adjust from an emphasis on communication to an emphasis on reception. Although regulatory and insurer requirements may be met by creating a trail of documents, physical or electronic, that tell workers about hazards, positive actual health and safety outcomes require the active engagement of workers in learning, self-protection and reflective practice. The professional needs to start by learning from the workers and developing change strategies that are consistent with their task performance and sensitive to the full range of constraints that affect this. This theme will be developed much further in the next two chapters.

## What needs to be considered when evaluating interventions and knowledge translation?

When evaluating the success (or not) of an intervention, its content will have an impact. In two of the case studies, the target audiences were supposed to be made aware of a new health and safety committee, and of a new electrical appliance policy. These did not require a direct or immediate change for the target audience, but an increased awareness of what needed to be done, either in relation to using the committee or in introducing a new appliance. In other case studies, more physical changes were being implemented. These could be visually checked: was hearing protection being worn or were face masks being used correctly? These latter interventions were both in smaller companies, where visual checks are more practical. Some companies used more informal routes for immediate feedback such as asking employees about new induction programmes or a redesigned intranet.

## Conclusions

This chapter has introduced a well-established way of thinking about the knowledge on which health and safety professionals base their practice. While this project was partly prompted by public criticism of the apparent lack of common sense in some health and safety interventions, it has also been an opportunity to reflect on the profession's approach to knowledge translation. Although the public criticism is often ill-informed, it is clear that there are problems in the process by which health and safety information and knowledge is translated from its original source through knowledge brokers and/ or boundary spanners and diffused to end users in the profession and beyond.

We began with a discussion of the characteristics of knowledge and its relation to practice. This emphasized its diverse, fluid and unstable properties. There is a considerable amount of work to be done in assessing its credibility, interpreting it for less specialist users and fitting its generalizations to specific contexts in managing local problems. Our research found a very wide range of potential sources that health and safety practitioners might use. This placed a considerable premium on their skills in identifying when and where to search for knowledge and in evaluating what they found.

It is certainly important to know which sources provide already-translated research knowledge, in the form of guidance and codes of practice. However, these sources cannot anticipate every possible problem and it is just as important, if not more so, to know what to do when these sources are not helpful and where to seek further advice. Nevertheless, most practitioners in this research were searching for guidance documents providing explicit facts and rules. While this underlines the need for these to be produced through robust, quality-assured processes, it seems that practitioners were looking for trusted sources at least in part as a substitute for having the resources or skills to make their own inquiries and come to their own judgements. Within the available pool of materials, health topics were less well covered than safety topics. This may reflect the more recent development of preventive efforts focussed on occupational health problems, which is noted in other chapters. While there is an expanding evidence base for practice in relation to health issues, this is still not fully translated into documents usable for practice. However, and this theme will be developed further in later chapters, there is a question about the balance between 'know how', understanding how to find and use knowledge to solve problems, and 'know what', knowledge of facts and rules for their usage, in the profession's approach (Senapathi 2011).

Some of this may be explained by the diversity of problems that practitioners are asked to deal with in their everyday work. Many have a specific background in one discipline – engineering, chemistry, medicine, psychology or whatever – but are expected to act as generalists. In such a situation, it is clear why they might choose to reach for a resource off the shelf (or the Internet) that can be applied in a straightforward fashion. Several of the projects, including this one, also picked up an undercurrent of professional concern about litigation, where the use of a recognized resource was a defensive practice. Following the letter of a guidance document would deflect a negligence action more easily than justifying a professional judgement. These drivers are in tension with the attempts to base regulation on principled and proportionate risk management rather than on rule-following.

When translating knowledge, consideration must also be given to the target audience in relation to the media used, languages and literacy level of the information. The reliance on guidance documents raised questions about the extent to which practitioners were equipped to think through the issues of reception. Although there have been considerable improvements in the presentation and readability of many of the products from knowledge providers, there often seems to be some uncertainty about the target audience: is this health and safety professionals as intermediaries or end users directly? In some cases, it was not clear how far practitioners saw themselves as having the responsibility, and needing the skills, for further translation as opposed to being a simple conduit for the message. Was the objective to provide information – and document for legal purposes that it had been provided - or was it to ensure that this information had been absorbed and acted upon by its recipients?

Both the survey and the case studies, however, did bring out the importance of the rich medium of face-to-face communication for knowledge translation in health and safety. This applied whether examining communications between health and safety professionals and a company's workforce or communication between researchers and health and safety professionals. Lean virtual communications were felt to be less effective: emails, for example, were not always read, and did not support two-way communication to check that knowledge had effectively been translated. Face-to-face communication, at least in principle, created opportunities for information recipients to push back and for practitioners to assess the effectiveness of their work in achieving engagement.

Health and safety professionals, as knowledge brokers and boundary spanners, need the skills to identify appropriate communication formats and the degree of translation required to match information on a topic to the capacities and concerns of their target audience. Not the least of these is an understanding of where the audience is located on Rogers's 'S' curve of adopters. The communication strategies for motivating early adopters are rather different from those for motivating laggards! The case studies also found that communication methods were associated with company size; both geographic dispersion in multiple plants, offices or outlets, and the actual number of end users. Knowledge translation is a particular challenge in some of the retail and hospitality organizations that are discussed in Chapter 6. These are often considered intrinsically low risk but, as our colleagues from Cranfield note, high-risk events still occur. An unsecured mirror may not be identified as a risk until it falls on a toddler – but a principled application of professional vision should have spotted it. The absorptive capacity of the audience and the prevailing safety culture are relevant to the acceptance of knowledge by company end users. Table 3.8 presents the skills required by the health and safety practitioner to achieve successful knowledge translation in relation to safety and health.

The case studies showed that effective interventions depend on a planning process that incorporates the health and safety knowledge broker's understanding of communication routes and translation. Once an intervention has been completed, it is also essential to evaluate its impact, either formally or informally, and feed this back into the organization to improve future practice.

What this work signals is a shift in the wider paradigm for professional practice in occupational health and safety. It might once have been argued that there was insufficient knowledge to enable sound health and safety practices to be implemented. This is clearly no longer the case. In fact, the problem is becoming much more to do with managing the quantity and relevance of health and safety information, and translating it from one form of knowledge to another. The knowledge translation approach is likely to become more important in future and the guiding frameworks and practical investigations

*Table 3.8* Collation of skills required by the health and safety practitioner

| Topic | Skills required by the OSH practitioner |
| --- | --- |
| Identification of knowledge | Search skills, ability to assess quality of knowledge. Understanding the risk appetite of the organization. |
| Persuasion | Understanding of the context of the intervention, face-to-face communication often better for new starts and training situations. Virtual contact may be required depending on the size and geography of the organization; consideration of how to evaluate this is vital. |
| Decision | Understanding that face-to-face interaction is often better. For larger organizations having a network of expertise across the organization to support decision-making can help. |
| Implementation | Ensuring that the employees are at the same level of readiness as those involved in implementing change – evaluation of safety culture can help with this process. |
| Confirmation | Vital to be able to evaluate whether the change has had an impact, through walkthroughs, risk assessments, observation, or other means of data collection including accident or incident rates. |

presented in this chapter represent a new and exciting frontier in the development of the health and safety disciplines. Indeed, it might be said that the field of knowledge translation itself requires more knowledge translation into the domain of occupational health and safety.

## Key points

- Knowledge comes in many forms. The professional challenge is to treat them with equal respect and to make appropriate choices about what kind of knowledge to use in what circumstances.
- Knowledge is fluid, dynamic and evolving. Health and safety professionals must always evaluate its credibility and relevance for their current practice.
- Guidance documents cannot cover all the contexts to which they might be relevant. Professionals must always be prepared to use their own judgement to adapt this knowledge to the particular workplaces and workgroups that they are trying to protect.
- Communication strategies should always start from the perspective of the intended recipient. How can knowledge best be absorbed rather than how can it be provided?

# 4 Health and safety knowledge in networked organisations

*Phil Bust, Alistair Gibb, Andy Dainty, Alistair Cheyne, Ruth Hartley, Jane Glover, Aoife Finneran, Roger Haslam and Patrick Waterson*

## Introduction

Occupational safety and health regulators and practitioners have traditionally tended to assume that they were dealing with strongly integrated organisations with clear, and relatively simple, lines of management and accountability. However, many modern workplaces and supply chains are actually networks, where groups of contractors and subcontractors are brought together to perform specific tasks within an overall plan. In such contexts, responsibility for health and safety, and the role of health and safety professionals, may be obscure and uncertain, given that this is often shared among multiple organisations and individuals. On the other hand, networks also create new opportunities to spread best practice as groups of workers come together, acquire knowledge and skills and then move on to take their places in new networks.

Our contribution to this research programme was to investigate the management of occupational health and safety in networked systems of production or service delivery. We compared practice in the healthcare, construction and logistics sectors. We aimed to identify the types of health and safety knowledge and evidence that circulate and work in relation to each other in organisations involved in networked delivery systems; how local actors in organisations interpret information; and how these combine with other influences to shape health and safety in practice. We sought to reveal ways in which health and safety knowledge, evidence and practices are produced, engaged and navigated, by means of observations of how organisations respond to formalised health and safety policy. We looked to reveal communities of health and safety practice, along with effective channels for the movement of knowledge, motivators and practices. In effect, we continued many of the themes relating to knowledge transfer discussed in the previous chapter with a more detailed focus on a set of contexts that had been identified as potentially problematic. A networked system may have a dominant partner with the resources to scan for knowledge and translate it into practice, which can then be imposed on other members. However, it may also be made up of a cluster of SMEs, none of which is individually well equipped to locate and use knowledge, let alone able to make demands of their collaborators.

We used an interdisciplinary research approach drawing on human factors, safety science, ethnography and organisation studies. We focussed on seven case studies across construction, healthcare and logistics, incorporating 150 face-to-face interactions in two phases, through interviews and focus groups. We developed data flow diagrams and used critical incident techniques to focus interviewees on practical realities. Emergent themes were then investigated further through qualitative ethnographic methods at selected case-study organisations from across the sectors, spending 5 weeks at each organisation working with 33 'key informants'.

In this chapter we explore health and safety knowledge flow, translation and enactment in various network and industrial contexts. This chapter focusses on data from the interviews and focus groups; other aspects of the research, including the ethnographic studies, are covered in our final IOSH report (Gibb et al., 2016a,b) and in Pink et al. (2013, 2014a,b).

## The 'in-between' mode of understanding, communicating and enacting occupational health and safety

Our work examined the sources of health and safety knowledge and the ways in which this was codified, built into organisational routines and procedures, translated into rules and finally interpreted and enacted. It focuses on the importance of looking at what people *do*, rather than looking solely at the formal rules.

We examined the routine practice of safety in workplaces and its relationship to the suppositions and expectations of formal rules, policies or statements about occupational health and safety. This approach responds to Hale and Borys' (2013) call for 'in-between' studies that reflect both the formal and the informal aspects of safety practice. Health and safety practice is something that arises in an interaction between the implementation of formalised rules and actual work, where formal and informal practices combine and interact in multiple and contingent ways. Our approach acknowledges the roles of both explicit and tacit knowledge – described in the previous chapter – revealing gaps between procedure and practice, and why these occur (following Dekker 2003). This also reflects the ways in which learning occurs in health and safety practice and organisation (see Gheradi and Nicolini 2002). An important implication of learning-in-practice is recognising that deviations from rules are not necessarily negative 'workarounds' or 'shortcuts', but necessary and practical adaptations based on the context of the task and the worker (see also Berlinger 2016).

## Networks and networked organisations

Our interdisciplinary approach enabled us to view the data through several different methodological lenses. It required us to grapple with, and challenge, traditional assumptions: does knowledge really 'flow', for example? We studied several types of networks and networked organisations, with differing implications for knowledge creation and knowledge flow. Case studies and illustrations are used to show how different network models or types can affect knowledge creation, flow and enactment.

We use the phrase 'networked organisations' to describe large, complex organisations where significant relationships and connections exist across traditional hierarchical, business or functional boundaries. They may be '*internal* where a large company has separate units acting as profit centres; *stable* where a central company outsources some work to others; or *dynamic* where a network integrator outsources heavily to other companies' (Business Dictionary 2015). Figure 4.1 illustrates the complex network of an NHS trust. Such organisations have been described as 'increasingly collaborative and knowledge-intensive' (Cross et al. 2007). Networks have a number of sub-networks which may be departments, divisions, functions, levels of hierarchy or locations of business units. They may also be different companies, acting together in consortia or as part of an integrated supply chain.

Viewing occupational health and safety across networked supply chains is key to understanding health and safety at a system level (Buckle et al. 2006). However, interrelationships between supply-chain processes often remain disconnected, causing disruptions to health and safety practices. Several researchers (e.g. Benjamin and White 2003; Walker 2005; Winkler 2006) have investigated the effect of fragmented supply chains on employee health and safety but they did not explicitly compare industrial sectors in a way that maps onto systematic differences in supply-chain configuration. This analysis

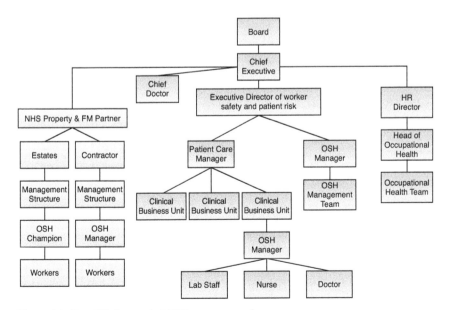

*Figure 4.1* Simplified generic NHS trust network structure

is important because promising practice in one sector may not fully apply in other sectors, given the complex and dynamic nature of networks. Walters and James (2009) investigated the role of supply chains in influencing health and safety at work and found that the internal dynamics of supply chains frequently lead to adverse outcomes.

However, our research found that some supply chains positively influence health and safety, particularly where there are external pressures from wider social, political and regulatory sources that create reputational risks. The precise effects of supply chains on health and safety can vary considerably even within the same sector. This can result from differences in such factors as the attitudes and objectives of buyers and suppliers, as well as the balance of power between actors in the supply chain. The term 'chain' suggests a more linear relationship than is often the case: 'supply network' may be a more appropriate description. We also found clear evidence for knowledge emerging through practice and not just as the result of networked effects.

## Findings

In this brief discussion of our findings we have chosen to focus on:

- where occupational health and safety messages come from (i.e. the sources of information used – both formal and informal);
- how information is translated and adapted as it moves through the network; and
- factors influencing health and safety enactment, specifically the worker end user; dynamic on-the-job risk assessments and parallel conflicting needs.

### *The use and translation of occupational health and safety policy and messages in networks*

Formalised health and safety policy is a key source of information in all our case studies. While the actual sources may differ, the main goal is to translate the information to match the business needs of the organisation. Here we discuss how this 'translation' occurs in practice. We investigate the key mechanisms and actors, along with the material and sensory contexts that shape knowledge creation, flow, translation and enactment. We also highlight the key characteristics of actors and influencers, both human and non-human, which facilitate these processes.

We asked participants about the best and least effective ways of finding out about occupational health and safety and about which sources they trusted the most and why. This complemented the survey work described in the previous chapter. Across all of the study organisations, health and safety information came from a number of sources and took several forms including *external* (coming from outside the network) or *internal* (coming from inside the network).

Conzola and Wogalter (2001) suggest that, given the same information, differences in the perceived characteristics of the source can influence the receiver's beliefs about the relevance of the information. In effect, as our colleagues from the *Institute of Occupational Medicine* (IOM) noted, information from a positive, familiar, credible, expert source is given greater attention. In the case of complex networked organisations there will be a primary, initial source, but also secondary sources as the knowledge flows around the network. Most actors in the network are both receivers of knowledge and subsequent sources of the translated knowledge for other actors, who they influence or for whom they have responsibility. This is also explored in our work on the construction of the London 2012 Olympic Park (Cheyne et al. 2012; Finneran et al. 2012).

In logistics, with a few exceptions, only health and safety managers and senior management used external sources of information on occupational health and safety. In construction, formal external information was normally only used by company directors, the health and safety director and managers. However, in healthcare, management-level ward-based staff also had access to external health and safety knowledge as part of their everyday work arrangements. Respondents said that this was due to the networked complexity of healthcare and the difficulty of facilitating training: time off from clinical tasks was a concern, particularly for contracted staff. Ward staff were given health and safety training and certification to train their colleagues.

Various external sources of information were mentioned across all sectors. Table 4.1 summarises these sources and estimates the prevalence across the interviewee types and across the three main industry sectors.[1] There was a hierarchy of information sources, with emphasis on the HSE (*Health and Safety Executive*), as the regulator, and IOSH, as the relevant professional body, for safety practitioners at least. This hierarchy essentially reproduces that reported by the wider IOM study in the previous chapter. In many cases, the information was 'pushed' from the source (e.g. a magazine or circular), but there was also evidence of where interviewees would seek out particular information.

Healthcare staff said that they obtained health and safety information through joint training sessions with colleagues from other trusts, often focussing on lessons learnt from implementation of a new procedure or process and stressing the benefits of the broader perspective that this brings. In construction, contractors shared health and safety information by discussing accidents on previous projects. Larger companies typically disseminated this learning across their organisations and construction sites, which included the site-based personnel of other companies in the network. It was hoped this information would then reach non-site-based parts of the subcontractors' sub-networks.

The HSE was the primary source used in logistics: both senior and junior managers felt it was appropriate to use information from the main regulatory body. IOSH outputs such as SHP (*Safety and Health Practitioner*) were also frequently cited by logistics health and safety managers, along with information from more specific organisations like the *Freight Transport Association*. Workers were found to use a wide variety of methods in obtaining health and

*Table 4.1* External information sources cited by interviewees

| External information sources | Citation frequency by interviewees | | | | | | | | |
|---|---|---|---|---|---|---|---|---|---|
| | Healthcare | | | Construction | | | Logistics | | |
| | *Worker* | *Manager* | *OSH* | *Worker* | *Manager* | *OSH* | *Worker* | *Manager* | *OSH* |
| Professional education | Most | Most | All | Most | Most | All | Few | Few | Few |
| Health and Safety Executive (HSE) | Few | Most | All | Some | All | All | Few | Most | Most |
| Other regulatory bodies[1] | Most | All | All | Some | All | All | None | Few | Few |
| Insurers | Few | Most | Most | Some | Some | Some | None | Few | Few |
| *Institution of Occupational Safety and Health* (IOSH) | None | None | All | Some | Some | All | Few | Most | Most |
| Professional bodies | All | All | All | Most | Most | All | None | Some | Some |
| Professional magazines | Some | Some | All | Some | Some | Some | None | Most | Most |
| Equipment/ product suppliers | Most | Most | Some | Few | Few | Some | None | None | None |
| The media (news about workplace accidents, etc.) | All | All | All | All | All | All | Few | Few | Few |
| Personal networks | Some | Some | All | All | All | All | Most | Most | Most |

1 E.g. the National House Building Council for construction or patient care bodies for healthcare.

safety information from external sources. These included actively searching for occupational health and safety information through the Internet and library sources; speaking to colleagues (past and present); receiving information alerts from health and safety agencies and indirectly from media sources (newspapers, television and radio).

Table 4.2 lists the main sources internal to the networks that were cited by interviewees.

Most of the networks studied had key individuals who acted either as 'central connectors' or 'brokers' (Cross et al. 2007). The central connectors are people who are frequently consulted for 'information, expertise or decision-making help'. Brokers are people who connect different subgroups in the network. Because of the networked complexity of each of the study organisations, certain workers were also given the opportunity to act as internal occupational health and safety knowledge hubs. Each healthcare ward, for example, had

*Table 4.2* Primary internal information sources cited by interviewees

| Internal information sources | Citation frequency by interviewees | | | | | | | | |
|---|---|---|---|---|---|---|---|---|---|
| | Healthcare | | | Construction | | | Logistics | | |
| | Worker | Manager | OSH | Worker | Manager | OSH | Worker | Manager | OSH |
| OSH managers | All | All | All | All | All | All | Some | Most | Most |
| Line managers | All | All | All | All | Most | All | Most | Most | Most |
| Colleagues | Some | Some | Some | All | All | All | Most | Most | Most |
| Acknowledged champions (keenies) | Few | Few | All | Most | Most | Most | Some | Few | Few |
| OSH Committee | Few | All | All | Most | Most | Most | Few | Most | Most |

workers who took on a health and safety role in addition to their own specific duties. They were provided with training to fulfil this role but, given the additional workload, they also needed a strong personal interest in health and safety to carry it out effectively.

In logistics, there were informal health and safety champions who, for one reason or another, had shown a keen interest in occupational health and safety, or were more experienced or happened to have had more training. These unofficial health and safety knowledge hubs were typically called the 'keenies'. Sometimes these people were overtly acknowledged and sometimes not. Construction workers also cited experienced co-workers as internal sources of occupational health and safety knowledge.

One healthcare trust had a person in every team acting as a 'kingpin' for health and safety, translating messages between the different layers in the network: management, medical and auxiliary. They would typically be a nurse or other healthcare worker who worked alongside the official health and safety professional and the line manager, acting as a secondary source for the ward staff directly. This approach had the advantage that the link person was close to the frontline workers and understood their situation and environment. It helped to ensure that messages were communicated appropriately.

In one logistics firm an occupational health and safety culture had been developed over a number of years. Workers knew that they were able to approach someone in the organisation who would be a source of health and safety knowledge and experience. These health and safety hub workers also provided a route for feedback from the front line to safety professionals.

Many of the locations visited had occupational health and safety committees or groups that generated information themselves or adapted generic information to suit the specific situation. The health and safety director from one healthcare trust stated that all the different divisions making up the trust network were represented at the occupational health and safety committee. He claimed that they would 'all happily and openly share with each other what they have been doing – what they have learnt'. While there were challenges in having a representative group, given the scale of trusts, with more than 2,500 staff, he still believed that the committee structure worked well.

The real source of health and safety information was not always clear. One logistics company had information from an occupational health and safety committee which was then channelled through line management or put on noticeboards, so the committee itself may not have been seen as the source. One interviewee did comment, though, that, as the pictures of the health and safety committee members were all on the noticeboard, it was easy to ask them directly for further information.

In healthcare and logistics, these committees also raised issues from the workforce and passed them on to 'management' for action. However, in some instances the excessive number of committees, especially in healthcare, that might require a say on a new initiative, meant that managers were more likely to find ways of working around them in order to introduce changes more quickly.

### Translation

The C-HIP model (Conzola and Wogalter, 2001) focusses on the process of translating messages in a personal sense: each individual who receives information from a source through a channel must then translate it before enacting it in their subsequent behaviour. According to C-HIP this translation is affected by the receiver's attention, comprehension, attitudes, beliefs and motivation. This section considers that individual, internal translation: what goes on internally to the individual receiver.

The cognitive abilities of the receiver and the psychological aspects of the internal processing of information were outside the scope of our project. Nevertheless, we did investigate barriers and facilitators that affected the taking in of the health and safety messages and their enactment. This expands on the notion of 'absorptive capacity', which was introduced in the previous chapter. The C-HIP model excludes other parts of communication theory such as 'noise' that affects the communication process, while our project included aspects other than direct, task-related, health and safety instruction to help answer the questions: 'how do you know how to do what you do and how to do it safely?'

One influence on the interpretation of health and safety messages was the worker's immediate environment. Healthcare participants, for example, noted that, while they were aware of appropriate procedures, they were often put in difficult positions where patient care seemed to conflict with their own occupational health and safety. Factors that hindered translation were also identified: for example, ambiguous messages which were not explained in terms of their consequences for a particular role. Lab-based staff at one of the case-study hospitals, for instance, said they were bombarded by hand washing posters, which really related to patient safety. The lab workers did not come into contact with patients. They also had their own hand washing protocols to follow as they worked in a lab (a sterile environment). Although these generic hand washing posters were not relevant to them, they were still posted in the lab.

The research also highlighted the need to check workers' understanding of the information they received, particularly where their first language was not English. Formalised checks of understanding were often taken a stage further by requiring workers to sign that they had heard or read and understood the message. Of course, this may not be very meaningful where the worker is not fluent in English and simply understands the signature as a necessary precondition for continuing to do the job.

The main action for most people is to pass health and safety messages on to others. This may occur in their own performance of a task, in co-ordinating their performance with that of others or in the course of duties as a manager or supervisor. Any individual must actually receive relevant information and process it internally, consciously or subconsciously, before passing it on, resulting in an inevitable element of translation as a message moves through the network. There may also be more deliberate adaptation of the message which may change its meaning in some way. It may be edited to make it more likely to be understood by the next people to receive it or to make it more palatable to them – 'to soften the blow' when they are asked to change an established way of working, for example.

In a logistics firm, information was interpreted by experts before being rewritten into a more practical form for site-based managers who, the experts thought, would not want to read 'reams of legislation and policy.' They acknowledged that the 'professionals' needed to know the legislation in order to check compliance: operational managers were concerned with delivery and more likely to read a one-page summary than a 50-page document.

The primary aim of healthcare is to ensure a high level of patient care, so task-specific health and safety knowledge was given where workers interacted with patients in order to protect both the patients and the workers. Where work on services (water, electricity, etc.) was required, for example, it was important to ensure that vital patient services were not cut off: arrangements were made to relocate patients, if urgent work was required and services needed to be disconnected. Patient dignity was also a consideration alongside staff safety when planning tasks. It was considered unethical to display information about patient illnesses where any passing member of the public could see it, for instance. However, domestic staff could not use the correct personal protective equipment without this knowledge, so a system of colour-coded stickers fixed to the door frames of the relevant rooms was introduced for their benefit.

In construction, a number of channels and channel characteristics were thought to aid understanding: practical training given by co-workers was seen as a good means of message transfer, for example. Conversely, there were problems with ambiguous messages, whose implications for a particular role were uncertain. Visual and physical aids were sometimes used to convey, or reinforce, messages: information about temporary walkways, for instance, was presented to employees through diagrams with colour coding to represent periodic changes in their location.

Managers and supervisors raised some difficulties in performing translation before passing the message on: a healthcare office manager, for example, considered that what she had been instructed to organise would not work 'further down the line', which then affected her enactment of the instruction. A number of interviewees also commented that the effectiveness of translation was often strongly influenced by the translator's own degree of commitment to occupational health and safety, which had an impact on their willingness to invest time and resources in the communication. Some managers might just 'go through the motions' to show that they had forwarded an item of information, where others would make a point of reframing the message and engaging their staff.

## Common themes

Some themes were consistently important: dynamic risk assessments; parallel needs; training needs; and the end user.

### The worker: the end user

In all the sectors, across the networks, frontline workers are much more likely to get hurt through doing their jobs. The real test for effective occupational health and safety knowledge flow is whether it gets to the frontline workers, as the end users, and whether they actually put it into practice.

Interviewees were asked to give reasons for why they behaved safely. The majority of responses related to a general sense of responsibility for safety which seems to permeate the workplace, including: caring for others; feeling responsible for other people's safety; and participating in the safety culture of the organisation. Despite the external barriers and enablers, a number of interviewees acknowledged the responsibility of individuals to listen, acknowledge, translate and enact health and safety carefully, effectively and appropriately. The culture and environment created in the networks and the individual workplaces was significant in encouraging or discouraging this.

One of the logistics organisations had a culture where everyone was encouraged to take responsibility for health and safety. It was apparent that there was devolved responsibility, both for oneself and for colleagues. This was shown in attitudes, beliefs and behaviours. The company had cultivated a 'caring culture' where management genuinely cared for the workforce, frequently identifying them as 'my' workers. Many workers felt able to approach colleagues if they perceived them to be behaving unsafely. The responsibility to 'do' something about unsafe acts was emphasised, although actually putting this into practice was seen to be more problematic. There was little evidence of 'us and them', with workers feeling able to raise issues with managers or senior managers: workers were listened to and actions were taken by managers as a result. Effective use of worker feedback was shown in one example where a task method had been chosen on health and safety grounds, but was then thought, by the workers, to be creating other health and safety problems – so the method was changed. There was evidence of both managers and workers striving to improve the business, with positive implications for profits and bonuses, driven by a belief that poor occupational health and safety would have a negative impact on the bottom line. Pragmatic reasons for preventing accidents reinforced genuine caring for co-workers' well-being.

The overwhelming view of increased personal ownership and responsibility was positive. However, in some cases it seems that the underlying culture was not supportive of this approach: workers were led to believe that, if anything went wrong, it would be their fault and they would get into trouble. As a result, they felt that the onus was on them to find the right guidance information to keep 'within the rules'. The importance of the role of the first-line supervisor in creating a supportive culture was stressed.

### Dynamic risk assessments

Dynamic risk assessments are partial enactments of the message because the individual considers, rightly or wrongly, that their conditions and environment make the message inappropriate, at least in part. Such *in situ* assessments – as our ethnographic research particularly enabled us to understand (Pink et al. 2014a,b) – were more likely where the situation was constantly changing, or more unpredictable, than in static, relatively unchanging work environments.

Terms such as 'shortcuts' or 'cutting corners' suggest that workers are not doing their job properly. They have a, usually negative, effect on the quality of the work or on health and safety. 'Making the rules work' often suggested the idea of pushing the limits as far as possible without actually breaking the rules – or interpreting the rules in a way that was probably different to their initial intention. 'Workarounds', however, were more often seen as pragmatic ways of doing a job that could not be performed exactly in the way that was initially planned. Workarounds would be the result of a conscious or subconscious dynamic risk assessment: in most cases the health and safety implications of the new method would have been considered before making the change. Workarounds were also more likely where the worker, rightly or wrongly, considered that the risk resulting from the new method was not great, probably because they considered that the risk from the original task was not great either. Thus, workarounds were considered less likely for high-risk tasks.

There is a clear difference between the environments faced in static, relatively unchanging workplaces and those where the situation is constantly changing or more unpredictable. In these less predictable situations, people often found they needed to assess the risk 'in the field'. For patient-facing staff in healthcare, the patient's condition might create uncertainty, whereas for community healthcare workers, or residential delivery drivers in logistics, it was the unpredictability of domestic settings. Healthcare staff training includes the use of scenarios, to provide the skills to cope with changing circumstances and to make dynamic risk assessments. However, *in situ* assessment of risk should not be used as an excuse not to plan or assess the risk in advance.

Home delivery workers in logistics faced the variable environments of the houses that they were delivering to. There were good examples of health and safety knowledge about special circumstances at a delivery point being passed between delivery workers. The first driver assessed the situation *in situ* and adjusted their delivery method, passing the information to future drivers so that they were better prepared.

Construction workers' environments change every day, but this is normal and is often, then, taken into account when risk assessments are drawn up. Moreover, in construction there is usually a manager or supervisor present, at least on larger projects. Individual workers have less need for discretion than a logistics delivery driver or community healthcare worker. In construction, true dynamic risk assessments are mainly required when unforeseen situations occur. Another area where *in situ* consideration of risk is needed is when a worker encounters a task or product where there does not seem to have been a risk assessment, or at least the worker is not aware of it.

In all of the sectors investigated there has been a move to spend more time considering unforeseen situations and to ask: what is the worst that could happen? Several organisations use mock-ups of emergencies to test the resilience of their systems and protocols and the ability of their staff to assess the risks *in situ* and respond appropriately, based on the established framework. Some tasks have legally prescribed restrictions but managers are advised not to try to control things that do not need to be controlled and to avoid 'making sackable offences out of trivia based on dubious statistical studies'. Our evidence suggests the best solution is to minimise the situations where managers try to control the risk by setting prescriptive rules and to emphasise training based on contingent, scenario-based approaches. Where tasks are recognised as high-risk, though, great care should be taken to avoid inappropriate alteration of the agreed method. Having said this, some high-risk situations are dynamic and those involved must have the skills and authority to assess the risk 'in the field' and take appropriate action.

Our research suggests that managers need to be more sophisticated in managing these *in situ* assessments. There is no 'one size fits all' approach that suits all sectors, network types, individuals or situations. There were differences between an individual worker deciding to use a workaround (which was not advisable) and a team discussing the issue and coming to the same conclusion, whether or not an official 'supervisor' was present and whether or not the workaround was subsequently acknowledged and adopted as standard (and by implication 'safe') practice.

Many contemporary occupational health and safety management systems stress the need for workers to be empowered to stop any task or process that they consider to be unsafe. For a small number of situations or tasks, though, 'stopping' the task is not an option: there could be a need for action to be taken immediately to prevent a catastrophe. Where such events are envisaged, individuals must be properly trained to make good decisions under extreme pressure. This is where scenario-based approaches are particularly valuable.

In many cases, particularly for lower-risk tasks, while a safe working protocol may have been agreed, there may actually be several alternative methods for doing the task safely. Moreover, the task environment may change such that the previously planned method is inappropriate. In such cases, an acknowledgement that the implementation of health and safety messages rarely matches the original source's exact intention can actually improve health and safety management and reduce the likelihood of unwanted consequences.

Workarounds will happen – they must be managed. Managers do not typically measure the success of workarounds – they only hear about the ones that fail. Practitioners should give more attention to the range of everyday (and often unnoticed) local adaptations of practice and health and safety guidelines that are intended to make work safer in *this* place at *this* time. It is normal for workers to fit 'formal' procedures to the contingencies of their job. This should not automatically be seen in a negative light. Adaptation and improvisation are not just inevitable: they may be *crucial* components of effective health and safety practice. They are opportunities for reflection and innovation that contributes to individual and organisational learning and growth.

Rather than trying to manage uncertainty by increasingly detailed and prescriptive regulation, it may be more productive to concentrate on supporting and enabling workers to become more skilful at improvising methods of achieving safety goals. There is a difference between health and safety as imagined and health and safety as actually done. If the workers have found a better way of working, that is a good thing. The challenge is to capture and transfer any good practice that emerges from these activities as a potential source for solutions that can be proposed elsewhere. Some workers certainly do take significant risks 'because they can', 'because it's quick' and 'because no one is watching'. However, many people are thoughtful, ingenious and attentive, using their cognitive and physical abilities to get the job done, and done safely. We should harness this behaviour as a resource rather than seeing it as a liability.

This perspective implies that managers and organisations need actively to seek deeper understanding of how and why such improvisations take place. On the basis of this understanding, they can determine the best ways to harness improvisation to produce safer working practices. Occupational health and safety guidance is, of course, necessary and vital. Given the impossibility of regulating for *every* scenario and future uncertainty, though, health and safety practice must be 'open' and 'flexible' enough to permit worker improvisation, while remaining precise enough to steer in the direction of safer working environments and practices. This remains the key challenge for those tasked with designing effective health and safety futures.

### Parallel (and often conflicting) needs

When we asked why workers do not always 'comply' with written procedures, we identified the problem of parallel needs, which conflicted with occupational health and safety requirements and expectations. Some of these, like time or cost pressures, were acknowledged to be unacceptable reasons for not working safely but were inescapable for some of our interviewees. Other needs were more nuanced, such as the safety or well-being of patients in healthcare.

### Time and cost conflicts

In construction and logistics, health and safety conflicts were largely related to performance and time pressures. Although workers felt that there was genuine concern for their safety on site, they also saw conflicts with time, planning and cost that led them to 'cut corners' or find 'workarounds'. These temptations were acknowledged but managers and supervisors often argued that workers knew that they must follow rules even if this meant a task took longer or became slightly more difficult. However, this message was not always supported by the contexts in which workers carried out their jobs.

Some workers were timed to do particular tasks: they believed this caused them to adjust the way they worked and lessened the priority given to health and safety. Limited storage space meant that some logistics workers had to work in confined and restricted spaces, especially during busy periods when stock needed to be moved in a short period. Warehouse workers handling pallet trucks explained that, at busy times, there were more trucks and hence less available space, leading to conflicts between health and safety and 'trying to get the percentage that they should be on'.

Busy periods were often mentioned as the time when 'shortcuts' would be taken: things 'don't always get done safely' but have to get done in order to meet targets. Healthcare interviewees explained that this conflict is exacerbated when wards are understaffed, due to cost constraints resulting from restrictions on healthcare budgets at local and national level. Construction projects are typically working to tight time schedules. The subcontracting system tends to pass time pressure down to the frontline workers. Many construction workers are still paid on the basis of the measured work that they actually do, which can create a conflict with health and safety.

### 'Patient safety' conflicts

'Patient safety' is a major factor that affects occupational health and safety knowledge flow, translation and enactment in healthcare. While the health and safety of healthcare staff is taken very seriously, a pragmatic balance is usually struck between patient risk and worker safety.

Healthcare staff would often choose patient safety over their own. Many healthcare workers consider their role as a practical, hands-on vocation rather than 'just a job' and feel passionately about their responsibility to care for the health and safety of their patients. Often, nurses would use the phrase 'my patient', illustrating their personal commitment. Healthcare workers are told that, if a patient is about to fall, they should guide the faller to the floor and not attempt to hold them up. However, nurses were described as being dedicated and caring, putting patient safety before their own, and some found this directive hard to adhere to. Pragmatically, staff also felt that patient injury would result in more serious repercussions than worker injury, such as a formal investigation into patient care. One ward nurse acknowledged the tension but explained that recent changes had started to correct this imbalance and put more emphasis on employee health and safety. In some healthcare situations, customised risk assessments allowed workers to make job and task adjustments to improve both the patient's safety and their own, ensuring that the correct tools and equipment are available for the job.

### 'The customer is always right'

Some logistics workers considered that the focus on performance for the customer meant that customer needs were sometimes put above occupational health and safety needs. An operations manager in a logistics organisation, which was having a particular campaign about increasing customer service, explained that drivers were 'ambassadors on the road'. Delivery standards had been established, including TVs being set up, washing machines connected and beds built. A delivery driver acknowledged that this increased customer service was important but that it did create conflicts. Nevertheless, our ethnographic studies suggested that delivery workers felt they had autonomy and institutional support to not put customer satisfaction above their own safety, to 'fail' a delivery if their health and safety were compromised.

## Conclusions

Our examination of the origins of occupational health and safety messages reveals a reliance on the regulator and other formal, professional sources for external information, with little variation across the three sectors. Internal information came from a range of sources, with health and safety, and line, managers playing a key role. The sectors varied, however, on how colleagues and particular health and safety champions came to be recognised and valued, either formally (in healthcare) or more informally, by experienced co-workers in all three sectors.

The interactions with unique working environments influenced how participants translated and enacted health and safety messages and knowledge, resulting in an emphasis on personal understanding and responsibility. The changing demands of these work environments made it necessary to devise the best approaches to safe working on the basis of local, *in situ*, assessments. These assessments are also influenced by the impact of parallel or conflicting needs, often related to the 'customer' (in the broadest sense) or to time and resource constraints. The combination of a dynamic work environment and parallel demands often results in improvisations. These should not automatically be regarded as negative but as a potential source of organisational learning and safety practice improvement.

## Key points

- Occupational health and safety practice is what people *do* and cannot be understood just by studying rules, policies or statements.
- The contemporary organisational context for health and safety practice is often a temporary network of relationships rather than a stable, integrated and continuing structure.
- Knowledge comes in many forms from many sources and is fluid, dynamic and evolving.
- There is an important difference between 'shortcuts' and 'workarounds'. The latter are opportunities to learn from the collective efforts of front-line employees to solve the problems that arise from the loose fit between any rule and the real world in which it has to operate.
- Health and safety interventions must be designed to fit with the competing pressures of other legitimate organisational goals such as customer or user safety and satisfaction.

## Note

1   The numbers of interviewees who mentioned each source have been grouped as follows: 100% = ALL; 51–99% = MOST; 11–50% = SOME; 1–10% = FEW; 0% = NONE. The sample size is not large enough for these groups to be considered significant but they are provided here to help present the overall picture.

# 5 Engagement of smaller organisations in occupational safety and health

*James Pinder, Alistair Gibb, Andy Dainty, Wendy Jones, Mike Fray, Ruth Hartley, Alistair Cheyne, Aoife Finneran, Jane Glover, Roger Haslam, Jennie Morgan, Sarah Pink, Patrick Waterson, Elaine Yolande Gosling and Phil Bust*

## Introduction

As previous chapters have noted, there is a particular area of concern about the impact of occupational health and safety issues on smaller businesses. Much of the criticism of 'red tape' that we met in the first two chapters focussed on the supposed problems for small businesses in complying with regulations that were too prescriptive and disproportionate to their available resources and the nature of the risks that they encountered. In the last two chapters, we noted the particular difficulties that small businesses experienced in connecting to knowledge flows. These might pass by because the smaller enterprise could not devote the same resources to scanning for new knowledge or absorbing it as might be possible for larger organisations. The intermediaries who brought the knowledge to the small organisation had their own biases, interests and agendas, which might not match those of their clients.

Small and medium-sized enterprises (SMEs) and micro enterprises are the majority of businesses around the world.[1] Despite their prevalence, though, comparatively little is known about how such organisations approach occupational safety and health. This gap is partly due to the practical difficulties involved in accessing these enterprises, which tend to be less 'visible' and harder to reach than larger organisations (Corr Willbourn 2009). However, it is also because SMEs and micro enterprises have traditionally been treated like miniature versions of large companies, neglecting their distinctive characteristics and contexts (Eakin and MacEachen 1998). Although there is an emerging body of literature on occupational health and safety in smaller organisations, this remains relatively limited and spread thinly across a wide range of sectors and geographical locations.

This chapter takes a fresh look at the way in which smaller organisations approach occupational health and safety. Our research (Gibb et al. 2016a) used a mixed-methods approach comprising 149 structured interviews, nine short-term ethnographies and 21 semi-structured interviews with owners and employees in SMEs and micro enterprises from a range of industry sectors, including logistics, healthcare and construction. This approach provided us with both breadth and depth of insights into occupational health and safety in SMEs and micro enterprises. It also allowed us to examine the views and experiences of employees, as well as those of business owners: previous studies have (often for practical reasons) tended to focus on owners and overlook employees.

We begin by discussing the key insights from previous studies of occupational health and safety in SMEs and micro enterprises. We then review the key findings from our own research and the chapter concludes with a discussion of the practical implications of our findings for health and safety policymakers and practitioners.

## Previous research

Previous studies have generally painted a negative picture of occupational health and safety in SMEs and micro enterprises – smaller organisations tend to be characterised as reactive and non-compliant, with low levels of knowledge and awareness. In their study of small manufacturing businesses in Sydney, Australia, for instance, Fonteyn et al. (1997: 54) found that 'the nature and extent of the owners' [OSH] knowledge was limited. Limited awareness and understanding of [OSH] legislation was a common problem'. This study found that one third of owners had no awareness of health and safety legislation: even those with a basic awareness did not understand their legal responsibilities. Similar issues were apparent in a cross-sector survey of small UK businesses by Vickers et al. (2005), revealing:

> a low level of awareness of specific health and safety legislation relevant to their businesses by respondents. Even in relatively high risk sectors, such as construction, only about half the respondents were able to broadly identify health and safety legislation that applied to their businesses.
>
> (p. 18)

Such findings beg the question as to whether knowledge of legislation and regulations necessarily means that people will comply with them and work in a healthy and safe way. However, a lack of knowledge and awareness of occupational health and safety requirements has been found to give rise to what is undoubtedly the most common theme in the literature on health and safety in smaller organisations: the tendency for owners and employees to under-estimate, discount or talk down health and safety risks and problems in their working environment. In a study of small food businesses in the UK, Fairman and Yapp (2004: 50) concluded that:

> Small businesses appear to lack the skill and knowledge necessary for them to be able to identify hazards within their premises. This leads to

confidence problems in identifying and rectifying problems. It can also lead to over-confidence and a belief that no hazards exist and that the public will not be exposed to food safety risks. This lack of knowledge contributes to the mistaken belief by many small businesses that they comply with the law.

Fairman and Yapp (2004) go on to claim that small business owners tended to determine their health and safety compliance purely on the basis of what a health and safety inspector had asked them to do, rather than from their own knowledge of legislative requirements. This echoes the findings of studies by Bradshaw et al. (2001) and Parker et al. (2007). Both suggest that small businesses tend to use a lack of accidents or incidents in the workplace as an indicator of their company's compliance.

A lack of resources – particularly time, money and information – has frequently been cited as a reason why smaller organisations appear to have low levels of occupational health and safety knowledge and awareness. Barbeau et al. (2004) argue that the 'realities of production' and the demands of keeping on top of day-to-day business mean that health and safety tends to be lower down the list of priorities than in larger organisations. Indeed, Champoux and Brun (2003: 16) suggest that:

> Some prevention and OHS management activities are practised regularly in small firms. The most common are activities that are also required to ensure production. However, activities that have a less direct impact on production (e.g. job rotation and the allocation of light tasks), and especially safety management activities, are much less common.

This links with the notion that, rather than being a formalised or structured process, occupational health and safety in smaller organisations is intrinsic or integral to the job or trade being undertaken – in other words, health and work are indistinguishable or inseparable. The ability to work safely is a reflection of the skills of the individual concerned (Eakin 1992; Corr Willbourn 2009). However, in some cases, this can lead to employees being blamed for accidents and injuries, because responsibility for health and safety has, in essence, been devolved to the worker (Holmes and Gifford 1997). It can also result in a fatalistic attitude to health and safety, in which accidents and injuries are viewed as an inevitable 'part of the job' (Holmes et al. 2000).

The literature provides an interesting insight into how social relations in smaller organisations can affect attitudes to health and safety. For instance, Eakin (1992) found that owner–managers said that they were reluctant to impose health and safety practices on workers because this would run counter to the norms of individual autonomy and non-hierarchical relationships in the workplace. They also wanted to avoid being paternalistic and, in some cases, felt that they lacked the authority to intervene to improve health and safety practices. Similar issues were identified by Parker et al. (2012: 474) who found that 'employers were conflicted about allowing employees a certain level of independence while also maintaining a safe workplace'. Research in Denmark by Hasle et al. (2012: 636) revealed that owners

> try to identify the standard that is generally accepted by colleagues, employees and authorities in the sector, and they try to be in line with that standard in order to portray themselves as decent people and protect themselves from personal guilt should something go wrong.

Hasle et al. (2009) argued that this desire for self-protection was also reflected in a tendency for owners to attribute accidents and incidents to 'bad luck' or unforeseeable circumstances beyond their control. However, employees might view these types of behaviour as an abdication of responsibility on the part of owners (Holmes and Gifford 1997).

Previous studies have shed light on the sources of occupational health and safety information in smaller organisations. A number of authors (e.g. Antonsson et al. 2002) have identified trusted intermediaries, such as insurers, accountants and trade associations, as important sources of health and safety knowledge, even though their importance is not necessarily supported by empirical evidence. Studies in the UK by James et al. (2004) and Fairman and Yapp (2004) both found that intermediaries were considered to be less important than health and safety inspectors as sources of knowledge. 'Informal' sources of knowledge, such as colleagues, business acquaintances and friends, have also been found to be important for smaller organisations. Research in the UK construction industry by Corr Willbourn (2009) revealed that respondents were more likely to listen to peers than the HSE and Brace et al. (2009) found that many micro enterprises said that they relied 'on newspapers, trade literature and builders merchants' for health and safety information. However, research in Australia by Fonteyn et al. (1997) raised concerns about the adequacy of such information sources. The importance of social networks is also brought out by the projects reported in the two previous chapters of this book.

Traditionally, researchers have focussed on understanding how (well) information is transferred from government and regulators to smaller businesses and, to a lesser extent, within organisations – that is to say, from owners and managers to workers, or vice versa. However, research in the UK construction industry by Corr Willbourn (2009), Brace et al. (2009) and Cheyne et al. (2012) found that health and safety information also flowed (or 'trickled down') from larger contractors to their subcontractors. There has been little research into whether (or how) 'trickle down' of health and safety information occurs in other sectors or whether the process works in reverse. We have seen in the previous chapters, for example, that experienced safety professionals would pick up the phone to each other and compare notes when they encountered problems but it is not clear how far this extends within their sectors. The flow of health and safety information between larger and smaller organisations is therefore potentially a missing (or at least poorly understood) link in the occupational health and safety literature. Furthermore, the question of whether health and safety knowledge and practice become embedded in organisations, once their formal relationships have ended, remains unanswered.

A number of studies have looked at how approaches to occupational health and safety compare in different sizes of organisations. Fairman and Yapp (2004: 50) suggested that 'small businesses have particular characteristics, and the process through which they make decisions as to whether to comply with legislative requirements will differ from those in larger businesses'. The literature suggests that larger organisations tend to be more proactive at dealing with problems and adopt more formalised processes for dealing with health and safety issues (Parker et al. 2007; Sørensen et al. 2007). Champoux and Brun (2003) attributed this to the fact the larger organisations tend to be more visible (to inspectors) and less isolated (better networked) than their smaller counterparts – resource availability was considered to be less of an issue. Nevertheless, there remains a lack of understanding of how or why approaches to health and safety change as organisations grow, or the way that knowledge and learning influence attitudes and approaches to health and safety.

With these questions in mind, our research set out to:

1   investigate the perceptions of occupational health and safety in SMEs and micro enterprises in the UK;
2   determine the sources of occupational health and safety knowledge in SMEs and micro enterprises;
3   identify the enablers and barriers to accessing and applying occupational health and safety knowledge in SMEs and micro enterprises;
4   examine how occupational health and safety knowledge is applied in practice in SMEs and micro enterprises; and
5   compare occupational health and safety knowledge and practices in SMEs and micro enterprises with those in larger organisations.

In this chapter we focus on discussing how, why and in what contexts, owners and employees of SMEs and micros learn, modify and communicate their *knowledge* about occupational health and safety in the workplace, and enact, or put into *practice*, this knowledge.

## Findings

### Occupational health and safety knowledge

A dominant theme in the literature was that SMEs and micro enterprises lack information, knowledge and awareness of legislative requirements and regulations. This finding was not borne out in our research. Although we did not set out to 'measure' or test levels of awareness amongst SMEs and micro enterprises, many of our research participants claimed to understand the regulations pertaining to their specific area of work, particularly in more highly regulated and/or high-risk sectors, such as mining and healthcare. This regulatory aware-ness, which seems to contradict the findings of previous studies, might be due to differences in study context, research methodology and/or the heterogene-ous nature of SMEs and micro enterprises. It may also be because:

- There is a perception that society (in the UK, at least) is becoming more litigious and occupational health and safety more regulated, meaning that SMEs and micro enterprises feel the need to keep abreast of regulations for peace of mind and reassurance.
- More widespread use of the Internet means that it is easier for SMEs and micro enterprises to find information about health and safety regulations and legislation in their area of work.
- There is a desire for SMEs and micro enterprises to be able to demonstrate their compliance with health and safety regulations in order to be able to undertake work for larger clients and companies.

The latter point was epitomised by an occupational health and safety con-sultant at a home warranty company, who mentioned that he received calls from micro construction companies anxious to know what they should do to comply with the rules on larger sites.

Examples such as this also contradict the findings of previous research (e.g. Antonsson et al. 2002; James et al. 2004), which suggested that SMEs and micro enterprises do not perceive the need for, or see the value in, new health and safety information. Although that was certainly the case amongst some of our participants, this was mostly because they felt that their working practices were already safe and that they had the required knowledge and experience to undertake their work safely. Owners and employees in smaller organisations seek new information when they feel that they have a need for it. For some people, the desire for new knowledge was about seeking reassurance that they are compliant – indeed, one sole trader in the construction industry suggested that tradespeople feel under pressure to attend health and safety training courses in order, he said, to indemnify themselves in the event of an accident. This desire for reassurance was closely linked with people feeling that they lacked knowledge of a particular aspect of health and safety, either because they were inexperienced or due to some external change, such as the introduction of new legislation or equipment. Clients or customers can also create a need for new information, particularly when the clients are larger organisations that have specific health and safety requirements. The owner of a haulage company described how, for certain jobs, he can be required to undertake (or attend) the same health and safety training that his client provides for its employees.

The nature and accessibility of the information available to SMEs and micro enterprises can be a barrier to seeking and acquiring new occupational health and safety knowledge. An employee in a small healthcare organisation described how she received longwinded emails 'full of a load of mumbo-jumbo' from the local authority and industry regulator, and that 'wading through' and 'deciphering' such information can be 'time-consuming and boring'. She felt that such information was 'not helpful' and often she did not understand it. Other interviewees described health and safety information as 'mind-boggling', 'confusing', 'tedious', 'contradictory', 'over-complex' and 'a bit anal', and pointed to the problems of 'jargon' and 'abbreviations'. In some cases, the information was not seen to be relevant to a person's particular (small business) situation. One healthcare worker (and microbusiness owner) commented on how the information provided by her professional body was 'very NHS biased' – relevant to managers of a hospital department but not really useful for her microbusiness. Elsewhere, the owner of a micro catering company explained how they received leaflets from their local authority notifying them of new legislation, but that she does not 'take a lot of notice of them' because they are not relevant to her business.

One way to understand how occupational health and safety knowledge circulates and operates in SMEs and micro enterprises is to identify where sources of knowledge are located. Tacit sources of knowledge – common sense and experience – were by far the most frequently cited, to the extent that many of the people we engaged with struggled to articulate how they knew how to work in a healthy and safe manner. They just knew, because it was obvious to them and it was what they did, day in, day out. This was particularly the case for people who had experience of carrying out the same job for many years and for whom working healthily and safely had become 'second nature' and 'just part of the job'. A farmer that we interviewed stated that health and safety was not something that he was consciously trying to achieve or knew how to do, but that he did things that were right and correct for him – things that made sense, such as how to drive his tractor or look after his animals. Elsewhere, an employee in a micro engineering company stated, 'How do I know? A lot of it's common sense. And if you've been using machinery and tools and things like that, cutting tools, you basically learn through a lifetime's work'.

Learning by doing was a particularly important source of health and safety knowledge for workers in SMEs and micro enterprises. It was closely linked with the notions of common sense and experience. The people that we engaged with explained how they had learnt from: their mistakes and near misses; by doing the same task many times; by observing and taking advice from others; and by solving problems encountered during the course of their work. The general sentiment was that learning was self-directed in smaller organisations.

Whilst tacit sources of knowledge were certainly seen to be important by participants in our study, we found that owners and employees in SMEs and micro enterprises used a wide variety of information sources, both formal and informal, and internal and external to their organisation – often in combination with each other. A person might use their experience, together with knowledge sourced from their colleagues and intermediaries or regulators. Industry bodies and associations were by far the most frequently cited intermediaries. They were generally seen to provide information in a form that was tailored to the needs of their members. The owner of a micro physiotherapy practice explained how her industry body had provided her business with advice on policies, disseminated articles on current issues and had set up an online forum that allowed her to exchange information with peers. Occupational health and safety consultants were also cited as a source of health and safety knowledge, although these tended to be used by SMEs and micro enterprises in more tightly regulated sectors, such as railways and mining. Employing external consultants was seen to be a cost-effective way for smaller businesses to access specialist and up-to-date health and safety knowledge not available 'in-house', reduce the administrative burden on managers and have reassurance that their business is compliant.

It also was evident from our research that SMEs and micro enterprises learn about health and safety from larger organisations, both formally and informally, in a number of sectors, including construction and logistics. This 'trickle down' tended to occur in situations where smaller companies subcontract to larger organisations, often as part of a supply chain or network. An occupational health and safety manager in a larger construction company described how his company had begun working with a smaller subcontractor that had a very low standard of health and safety, but within a year the subcontractor had won a 'best at health and safety award'. The manager explained how the subcontractor had been given access to training and assistance in developing their health and safety programme, allowing the subcontractor to understand what was needed without fundamentally changing the way they ran their company. One particularly interesting insight from our research is the way in which people carry knowledge with them from previous employment(s), which was often with larger organisations such as the NHS – more often than not the knowledge had been gained through formal training. In some instances, people carried health and safety knowledge with them from a different area of work and then applied it to their new area of business. Two people that we interviewed had previously received manual handling training when working for large multinational companies and now applied that knowledge in their current jobs as market traders. This suggests an indirect 'trickle down' of occupational health and safety knowledge from larger to smaller organisations.

### Occupational health and safety practices

The way in which owners and employees in smaller organisations enact, or put into practice, their health and safety knowledge has actually received very little attention in the literature, perhaps because of the practical and ethical issues involved in exploring such practices. We tried to overcome this by engaging (where possible) with participants in their place of work, providing opportunities for participants to 'tell' as well as 'show' us how they 'knew how' to work in safe and healthy ways. We found four main types of enactments: ones that involved gathering information; ones that involved sharing information with others; ones that involved doing something; and ones that involved avoiding doing something. Some of these enactments were relational – that is to say, they involved other people. Others were individual. In some situations, actors may take multiple steps to maintain a healthy and safe working environment. The owner or manager of a business might: observe her employees; tell them if they are doing something unsafely; demonstrate how the task can be undertaken safely; and then observe her employees again to confirm that they have adopted the safer practices.

Personal protective equipment was the most commonly mentioned enabler of healthy and safe working across a range of sectors, although some interviewees suggested that they used their personal judgement about when to use such equipment – being mindful of the task in front of them and the most appropriate protective equipment required to complete that task safely. There were instances of people describing items of protective equipment as 'restrictive' or 'unnecessary' in some situations. A chimney sweep interviewee said that he does not like wearing a hard hat because it gets in the way of him being able to see properly, particularly when working in small, enclosed spaces. However, he had to balance that with the risk of suffering a head injury. Elsewhere, a gardener described how the guards on his strimmer were too close to the strimmer head, which meant that it was difficult to use – he would therefore move the guards upwards a little bit and compensate by being more aware of people being around him.

Our research highlights how owners and employees in SMEs and micro enterprises have developed working practices that, in some cases, might not be in line with formal recognised practice, but that nevertheless seem to be safe within the context that they are being applied. Knowing how to work in healthy and safe ways is generated from the interaction between people and the specific social, material, sensory, affective and regulatory contingencies of the workplace environments through which practitioners undertake practical activity. This was particularly evident in the insights provided by the short-term ethnography. For instance, removals workers described 'knowing how' to pack, lift and transport boxes through bodily sensation; a mobile beauty therapist spoke about relying on 'gut feeling' to make decisions about whether homes are safe (or not) to work in; and a company administrator described paying attention to cues like tension in her body to judge when she should take a short break from computer-based work. Although evident in different sectors, these ways of knowing were not usually articulated by participants because of their routine and taken-for-granted status. For workers in smaller organisations, these 'other' ways of knowing become part of the everyday enactments that they perform to do their work safely. 'Safe improvisations' are their adaptive responses to the varied workplace environments that they encounter.

In contrast to workers in larger companies, workers in our SMEs and micro enterprises did not appear to regard occupational health and safety as something 'owned' by an organisation. Their reflections on their working practices suggested that it was less easy for them to separate health and safety from their individual responsibility and jurisdiction. It appears that health and safety converges through, and becomes 'internal' to, practitioners themselves and is expressed through their everyday routines and working practices. Occupational health and safety practices in SMEs and micro enterprises are (more often) located at the individual rather than the organisational level. They are bound up with a broader notion of 'taking care' of oneself and/or (what could be characterised as) being a 'responsible', 'committed' and 'competent' practitioner. At an individual level, a fear of being hurt or injured was the most frequently cited motive for taking health and safety seriously in the workplace. Working in a healthy and safe manner was about 'self-preservation', 'looking after yourself' and 'wanting to go home safe'. People tended to explain their fears and concerns by making reference to specific hazards arising from their work, such as falling from a roof or lifting a patient. For some people, concerns for their health and well-being were reinforced by the knowledge that being injured in the workplace could jeopardise their livelihood – a particularly important issue for sole traders and smaller micro organisations, for whom being unable to work would mean lost income.

Peoples' desire to work in a healthy and safe manner was also motivated by concerns for the well-being of others – colleagues, employees, customers or members of the public. Indeed, for some people, the fear of hurting or injuring someone else was their primary motive for healthy and safe working, such that they would be more accepting of risks to themselves than they would be for others. Family relationships in some micro enterprises also meant that individuals took a greater level of care, perhaps more so than if they did not know the people that they were working with (as is often the case in large organisations). Concerns for the safety of others was also partly about the fear of being prosecuted or sued (particularly by members of the public) in what is seen to be becoming a much more litigious society. Taking health and safety seriously was therefore a way for sole traders and small business owners to gain reassurance and peace of mind that they were compliant. However, it was also about peoples' pride in their work and a genuine desire to operate professionally and responsibly – not just fulfilling a legal duty of care, but a moral responsibility to do the right thing. For some people, working in a healthy and safe manner was more about a sense of personal responsibility ('I wouldn't do anything different than what I'd do for my mother'); for others, it was very much a feeling of professional responsibility ('I'm a nurse. You don't take risks').

The working environment was the most frequently mentioned barrier to healthy and safe working, particularly for workers who are 'out in the field' and have less control over their working environment, such as lorry drivers, domestic tradespeople or healthcare workers who visit peoples' homes. In these contexts, the ability of (and necessity for) workers to adapt and improvise towards safety is perhaps especially heightened. In many of these situations, the standard of working environment may be determined by the client or customer. One construction sole trader described how sometimes, when subcontracting for another company, he might not be provided with the proper scaffolding or platforms to work on. A fitness instructor explained how she switched to a new venue because the building she had previously hired for her classes was dirty and unsafe – several times she had found nails and screws on the floor. However, in some cases, the reasons for changing working practices were more subtle and proactive: not necessarily a specific incident or near miss, rather a progressive realisation that a particular practice is not working. This might be due to a deteriorating health problem – a bad back that creates discomfort or prevents someone from doing a particular task – or just the sense that a particular activity is 'hard work' or unsafe.

Changes to working practices were also triggered by external factors, particularly client requirements and new regulations. A number of interviewees suggested legislative changes would be the only reason why they would make changes to their working practices, because then they would have no choice but to take action. However, in some cases, small businesses argued that they were unable to comply with new regulations and adopted alternative practices to work around the legislation instead. A contract gardener described how the restrictions on the storage of pesticides mean that he now uses different methods, even though these are more time-consuming and less efficient. Clients were also seen to be an important trigger of changing occupational health and safety practices, particularly in situations where smaller businesses were working or subcontracting for a larger client. One sole trader working in the construction industry explained that he uses 110 V power tools on larger building sites, because that is what his clients require. When working for domestic clients, he uses 240 V power tools because he feels that he can work safely with them and that 110 V tools are unnecessary.

Within the SME and micro sector itself, it was evident that organisations with more employees tended to adopt more overt formal occupational health and safety processes, and more formal channels of communicating health and safety knowledge, although tacit knowledge creation and flow was evident in all the organisations. While our data does not allow us to pinpoint a specific 'tipping point' for when overt occupational health and safety becomes more formalised in SMEs and micro enterprises, it was clear that the move from sole trader to micro organisation is a significant step, because of the sense of responsibility that comes with employing other people. Indeed, this was a more noticeable tipping point than the one that comes with employing five or more people which is an important cut-off point in legal terms in the UK (because organisations with fewer than five employees do not need to have a written health and safety policy or to record their risk assessments). The level of formalisation of health and safety practices was also influenced by the type of work being undertaken. Even micro companies working within more regulated and/or hazardous sectors, such as mining, used more formal processes, whereas similar sized organisations in less hazardous and/or regulated sectors relied much more on informal processes to maintain safe working. These findings underline the fact that SMEs and micros are not homogenous. It is important to recognise the differences that exist between them, due to differences in organisational culture, the type of work being undertaken and the sector that an organisation operates in.

## Conclusion

We drew three main conclusions from our research into occupational health and safety in SMEs and micro enterprises. First, while the literature generally paints a negative picture of how SMEs and micro enterprises perceive and deal with health and safety, our findings paint a more positive picture: many of our enterprises recognised, for varied reasons, the importance of health and safety in the workplace. Some participants were frustrated by specific rules and regulations, but this was less about rejecting health and safety *per se* than with a feeling that the rules and regulations were inappropriate to their context, creating unnecessary bureaucracy and, in some cases, making the workplace less safe. Many of our participants thought working in a healthy and safe manner was just the responsible thing to do, an intrinsic part of their work, and a key aspect of operating their business. It is difficult to explain the discrepancy with previous studies: it may be due to differences in context and timing, research methodology and/or the heterogeneous nature of SMEs and micro enterprises.

Second, our findings suggest that SMEs and micro enterprises use a wide variety of formal and informal sources of occupational health and safety information, often in combination with each other. Tacit ways of knowing, drawing heavily on common sense and experience were particularly important and trusted sources of knowledge for owners and employees in small and micro enterprises. There is evidence to suggest that some SMEs and micro enterprises benefit from 'trickle down' of knowledge from larger organisations. Owners and employees carry health and safety knowledge with them from previous jobs, including those with larger organisations. We suggest that this is an important, and frequently overlooked, source of health and safety knowledge in the SME and micro sector. It also underlines the fact that owners, managers and employees in smaller organisations do not just passively receive information – they use the information to create, shape and adapt knowledge through their everyday practices and interactions with other actors in the workplace.

Finally, our research has highlighted how owners and employees in smaller organisations have developed working practices that, in some cases, might not be in line with formal recognised practice, but nevertheless appear to be safe within the context where they are applied. There is far more to enacting good, effective occupational health and safety than mere compliance – even if the rules were the best and most appropriate rules that could be. Personal, tacit ways of knowing should not be assumed to be incompatible with formalised health and safety. It would be more productive to acknowledge, and seek to better understand, the ways that these become complementary and the ways that they do not. Workers in the small companies we studied skilfully blended diverse ways of knowing. In most cases, this led them to perform their work in general compliance with regulated health and safety, while attuning their practice to the contingencies of varied workplace scenarios and environments. For workers in smaller organisations, these 'other' ways of knowing are part of the everyday enactments that mean they perform their work safely. 'Safe improvisations' help them adapt to the varied workplace environments that they encounter. Such practices are more reflective of a 'Safety-II' perspective that 'explicitly assumes that systems work because people are able to adjust what they do to match the conditions of work' (Hollnagel 2014: 137). However, they have received little attention in the literature on SMEs and micro enterprises. Where they have been noticed, researchers have tended to view such adaptations as risky and dangerous rather than as creative responses to the context that may be a basis for improvement and learning.

The challenge for health and safety practitioners and legislators is to bring together different ways of knowing and enacting occupational health and safety (including both the regulated and the tacit), and then to design ways to better support workers in this complex process. This should include helping workers in smaller organisations to make judgements, responses and adaptations towards safety outcomes. While we accept that there is a clear need for some formally codified health and safety guidance (especially in high-risk work settings), there is also a need to acknowledge the diverse ecology of knowing and practising health and safety in workplaces. Workers may be 'nudged' towards safer and healthier practice. Rather than focussing exclusively on how occupational health and safety can be improved through more comprehensive or tighter regulations, it may be more productive to understand the myriad ways in which workers *already* do their work safely. Empirical studies of how people learn, share and use health and safety knowledge can identify effective practice, while also understanding why there are gaps between formalised health and safety and everyday practice. From these studies, practical interventions associated with training, communicating and regulating safe working may be developed. However, such interventions need to reflect the heterogeneity of smaller organisations. Health and safety practices are more contingent and varied for smaller organisations than they are for more highly regulated, scrutinised and routinised larger organisations.

## Key points

- SMEs and microbusinesses are not just scaled-down versions of large organisations. There are important qualitative differences in their culture and internal relationships that have an impact on occupational health and safety.
- Owners and managers generally seem committed to health and safety objectives but have limited access to the knowledge required to achieve these. 'Trickle down' from collaborations with larger organisations is an important source for practice improvements.

- Owners and managers can be reluctant to enforce safe working practices on experienced employees. Health and safety is more a matter of individual than organisational responsibility.
- Regulatory approaches often tacitly assume resources and structures that are absent from organisations at this scale. 'Workarounds' need to be examined in a more positive light as genuine attempts to achieve health and safety objectives by complying with the spirit rather than the letter of regulations.

## Note

1  In this chapter, we use the *European Commission*'s (2003) definition of micro, small and medium-sized enterprises as organisations employing less than 10, 50 and 250 people, respectively.

# 6 Safety leadership: fashion, function, future

*Colin Pilbeam, Noeleen Doherty and David Denyer*

## Introduction

For more than two decades, stakeholders in the safety field have promoted the idea of safety leadership. Researchers have investigated the personal traits or behaviours that are supposed to encourage it. Professional bodies have urged individuals to enact it or suggested ways in which it can be stimulated. Policymakers and accident investigators have demanded that organizations show more of it. This continued attention to safety leadership suggests both that it is important, and that it is difficult to achieve. In the final substantive chapter of this book, we will discuss the challenge of taking forward the findings reported by the other projects about the need to manage the context in which health and safety interventions are developed and delivered. This context has several dimensions, which have been explored in previous chapters: the legitimacy of intervention at the level of a whole society and its government; the flow of knowledge from researchers into organizations and networks; and the implementation of better practices in the routine performance of tasks in the workplace. Leadership cuts across all of these dimensions. While it is not essential that every health and safety professional operates at each level, it is essential that they understand the processes involved and their implications for the particular level where they are trying to have an impact.

This chapter begins by reviewing the general challenges that face anyone who is trying to exercise leadership in health and safety. We will then introduce some of the relevant findings from previous research – a clear problem with the concept of 'safety leadership' is its weak connection with the substantial body of work on leadership that has been conducted in organization studies. Finally, we will introduce our own work from a study of supposedly low-hazard organizations and the issues that they present for people who are trying to give a lead on occupational health and safety issues.

## Why is safety leadership so difficult?

Safety is primarily about prevention. This objective means that practitioners tend to follow a 'defensive agenda' to maintain safe working conditions and ensure the continued safety of personnel. This contrasts with the 'progressive agenda' of other managers and professionals, which seeks to innovate and make new things happen (Buchanan and Denyer 2015). The different characteristics of these agendas frame the difficulties of enacting safety leadership.

First, according to Karl Weick, the American social psychologist who has spent time investigating the development of high reliability organizations (e.g. Weick and Roberts 1993), safety is a 'dynamic non-event' (Weick 1987). Nothing should happen. Inevitably this is unexciting or as the ACSNI report noted, 'there is no doubt that safety is often perceived as boring' (HSE 1993: section 154). It is not easy to encourage individuals to embrace the safety challenges of an organization with enthusiasm. This is much less interesting, and may have less career value, than driving other organizational agendas, about innovation, quality or efficiency, that can be identified with their author.

Second, accidents and incidents may be seen as atypical or uncommon. Any subsequent interventions are, therefore, likely to be questioned and resisted as potentially unnecessary and costly overreactions. This response is particularly likely in service sector organizations. These employ the majority of UK workers but rarely experience events that are likely to result in serious or life-changing injuries.

Third, safety agendas may arise from recommendations made by a safety review, perhaps following an incident. These recommendations may be felt to be imposed from outside, and, to that extent, resented by those affected. They may be seen as interfering with existing working practices, particularly if they add layers of processes, and create circumstances where 'it is impossible to make an unproblematic choice because of contradictory rules' (Mascini 2005: 482).

Fourth, and related to the previous characteristic, is the way the defensive agenda is constrained by statutory regulatory requirements. These reduce the freedom of individuals and organizations to be creative and innovative: the changes deployed may, therefore, be resented or resisted. They may appear to have been embraced – but with reluctance because there are too many other important things to do – so the agenda is not prioritized.

Finally, the agenda may be rejected because stakeholders have other opinions about the nature of the hazard and the degree of risk. These views lead the stakeholders to question the legitimacy of the actions intended to minimize the hazard or risk.

Safety leadership demands the development of a defensive agenda. This, however, seems to stimulate a set of defensive responses among those affected by the actions required to deliver the organizational safety agenda. It is not, then, surprising that safety leadership is difficult to enact. This difficulty is compounded by the absence of a clear definition of safety leadership and, with that, clarity over who is a safety leader.

## What is safety leadership?

At a fundamental level, safety leadership has never been defined. Every stakeholder may have a different perspective on what safety leadership is, who is involved, and what they need to do. The available definitions are typically driven from the bottom up by whoever happens to be involved in a particular study, and what they do. While this is a pragmatic response, it lacks conceptual clarity (Suddaby 2010), impeding the accumulation of knowledge, making comparative studies impossible and the development of practical guidelines difficult.

The primary aim of safety leadership is the prevention of (non-trivial) accidents or injuries. This may be achieved by a measurable reduction in the number of accidents or injuries in the workplace and/or an observable change in the behaviour of employees so that both compliance with, and participation in, safety measures are improved. It is relatively easy to demonstrate success in achieving the first aim by reference to appropriate statistics. Hazard reductions can be achieved by redesigning work processes; appropriate training and supervision; ensuring adherence to standard operating procedures; and providing appropriate personal protective equipment (PPE). The second aim is more challenging because individual behaviours respond to a variety of different stimuli. Safety behaviours are commonly explained in terms of either extrinsic or intrinsic motivation. Extrinsic motivation, based on reward (which includes praise and feedback as an alternative to tangible rewards), is assumed to encourage compliance (Dahl and Olsen 2013). Participation, on the other hand, is promoted by factors such as stimulation and challenge that trigger intrinsic motivation in the individual employee (Kapp 2012a).

Safety leadership has a specific focus – safety. Unlike other specialist roles that have acquired leadership status, like IT leadership or sales leadership, however, safety is not functionally-based and its performance target is rarely set in financial terms. As a consequence, safety may not be seen as integral to the performance of the organization. Managers do not give it priority, and emphasize it to other employees, even though they have 'prime responsibility for accident (and ill health) prevention' (HSE 1997). Safety is often seen as a cost rather than as a saving, ignoring the adage, 'if you think safety is costly, try an accident'. By identifying safety leadership as a distinct activity, safety in organizations is marginalized, and possibly separated from mainstream activity, rather than being integrated into everything an organization does. This separation encourages it to be seen as a cost rather than contributing to the efficiency of the organizational processes and increasing margin (see for example, Evans et al. 2005).

The notion of safety leadership raises two particular questions. Why does leadership in general not embrace safety adequately – is safety leadership different from leadership more broadly? Alternatively, if leadership does embrace safety, why does it not do so adequately, so that a separate and distinct activity is necessary? Part of the answer to these questions may lie in the perspectives used to explore leadership in general and safety leadership in particular. As Ladkin (2010: 2) noted, the 'one thing that is clear about the leadership literature is that there is relatively little that is clear about leadership'. If the primary concept – leadership – is poorly defined, it should not surprise us that safety leadership is equally ill-defined. Much of the literature that investigates leadership employs research methods that seek to identify particular leadership characteristics, traits and competencies (e.g. Gordon 2002). Investigations of safety leadership have followed a similar path. However, this approach focusses on an individually-based unit of analysis – the *leader*. It ignores the collective processes of *leadership* which encompass the leader, their followers and the details of the particular context in which they interact. By focusing on particular characteristics or traits of the individual, or on their interaction with another individual, leadership studies (including those of safety leadership) fail either to recognize the meaning attributed by the leader (or others) to the particular activity or outcome, or to judge its relative success, which depends on context (Ladkin 2010). Leadership in general, and safety leadership in particular, need to be more broadly conceptualized.

Grint (2000) argues that leadership can be summarized as a portfolio of four arts (The Philosophical Art, The Fine Art, The Martial Art and The Performing Art) concerned with establishing and coordinating the relationships between four things: the why, the what, the how and the who. These suggest that leadership has four purposes: to provide meaning; to give direction; to prioritize activity and to build community. Safety is implicit within each of these. Safety leadership should not, then, be separated from leadership more generally. It is difficult to imagine leadership that provides meaning, gives direction, prioritizes activities or builds community, while disregarding the health and safety of the individuals concerned with these things.

## Who are the safety leaders? Suggestions from the literature

As with the wider field of leadership studies, the focus of safety leadership research is typically at the level of individuals. Most research studies have focused on supervisors and team leaders (e.g. Zohar and Luria 2003; Conchie and Donald 2009) and only very occasionally looked at senior managers or board members (Smallman and John 2001). However, much of the specific guidance on safety leadership from the *Health and Safety Executive* relates to the senior management team or board members (e.g. HSE 2004) and less frequently to supervisors and team leaders. Health and safety legislation however takes a more inclusive view. Rather than emphasizing one group over another, it highlights the need for everyone to attend to health and safety and to show concern for their fellow worker(s) (*Health and Safety at Work etc. Act,* 1974: section 7), although the law does require greater accountability for safety from directors and senior managers.

While individuals are most often conceived of as the safety leader, whole industries or sectors have sometimes been encouraged to demonstrate safety leadership. The Cullen Report (2001: section 5.12) on the Ladbroke Grove rail disaster noted that 'there was no clear identification of safety leadership in the UK rail industry', suggesting that safety leadership applies across industrial sectors as well as within individual companies in the sector. If the term 'safety leader' can be applied to specific individuals, to work groups (e.g. senior management teams) or to entire industries, the risk of confusion over its meaning and usage should be obvious.

## Who are the safety leaders? An empirical investigation of service sector organizations

Our study examined safety leadership in 11 retail outlets from four national retail chains, and in two warehouses and a central office of a global logistics company (Pilbeam et al. 2015). More than three quarters of the people we interviewed in these service sector organizations identified themselves as being personally responsible for safety. One retail worker noted that, 'it says on the board you are all responsible'. Typically, this involved 'following the rules', or more colloquially, 'doing nothing stupid' or 'trying to keep out of mischief'. For most, 'following the rules' meant wearing PPE, tidying up, lifting correctly, challenging others and looking out for potential safety issues. For others, it meant participation in health and safety meetings and training others who were less experienced. For managers, it meant leading by example or acting as a role model, as well as completing the periodic audit assessment sheet.

However, more than half of the interviewees both identified everyone as responsible for safety *and* a specific person or groups of persons as responsible for safety. The balance between these two perspectives varied by sector. In an office environment, three quarters of the interviewees identified a single individual or the safety committee as having responsibility for safety, but only one quarter suggested that everybody did. It was either 'down to the individuals to be aware of their own safety' or 'absolutely nobody' had responsibility. In the retail environment, about 60 per cent of interviewees said that managers and supervisors or the health and safety representative/champion were responsible for safety. A similar proportion noted that it was down to 'everybody really', and that 'we're all responsible for making sure that everybody does things in the correct way', although 'not everyone cares'. This suggests that universal participation in safe working may happen more 'in theory' than in practice. In the warehouse, only 30 per cent of interviewees identified the manager or health and safety adviser as being responsible for safety, while 70 per cent thought that 'everybody does their bit', and all were responsible for each other', although 'some don't bother'.

In summary, both particular individuals and everyone were responsible for safety in the organization and therefore were required to demonstrate a degree of safety leadership.

### Genesis of the term – safety leader

The third edition of the HSE publication *Managing for Health and Safety –* HSG65 (HSE 2013) identifies both *leaders* and *managers* as key actors in *managing* health and safety. The publication draws clear distinctions between the activities associated with each role as they deploy the suggested Plan, Do, Check, Act approach to safety management. Leaders are not necessarily line managers: in this guidance they are typically senior managers, directors or members of the board. This publication also refers to *leadership*, which, together with training and involvement of workers, is 'key to effectively managing for health and safety' (HSE 2013: 10). A set of leadership processes are identified, but then linked not to leading but to managing. This is, to say the least, confusing.

One way to clarify the concept is to look at the way references to safety leadership have evolved over the last 25 years, and to ask what the authors seem to have meant at the time. Figure 6.1 shows a timeline of the publication of various reports or guidance notes by the HSE or HSC, different standards published by the *British Standards Institution* and several accident investigation reports from serious incidents indicating who the focal actor was and what they were supposed to do, whether leading or managing safety. The figure also tracks the appearance of the term 'safety leader' or 'safety leadership' in these literatures, and to whom this was applied. Before 2000, people mostly wrote about managing safety. Since 2000, safety leadership has become a more common term. However, until very recently, this was applied mainly to the roles and responsibilities of directors or senior managers. It is only in the last few years that everyone – in the research literature, supervisors in particular – has been expected to demonstrate safety leadership.

The ethos of the *Health and Safety at Work etc. Act* 1974 was 'command and control'. This was reinforced by the HSE leaflet, *It's your job to manage safety* (HSE 1991a). This was aimed at directors. It drew on comments from the then recent disasters (including: *Herald of Free Enterprise* [Department of Transport 1987], *King's Cross Underground Fire* [Department of Transport 1988]) to point out the legal duties of directors towards safety enshrined in the 1974 Act. However, it notes that 'safety is a management objective' and that 'poor health and safety performance is a reflection of bad management', suggesting that directors are expected to manage safety directly.

At the same time, the first edition of HSE's guidance document *Successful Health and Safety Management* (HSG65) (HSE 1991b) was published. In the 'POPMAR' management framework included in this guidance, 'visible and active leadership of senior managers is necessary to develop and maintain a culture supportive of health and safety management'. This summarized the guidance on organizing but without any further elaboration in the rest of the document. Senior managers were to establish an organizational culture conducive to safe working and so indirectly manage safety.

This emerging orientation towards safety leadership was overtaken by *The Management of Health and Safety at Work Regulation* (1992), needed to implement an EU council directive (CEC 1989). This shifted the focus from leadership back to managing health and safety, which was affirmed in BS8800 (BSI 1996). This direction continued for several years: accident reports of the time emphasized safety management, but did not mention leaders or leadership (e.g. *The fire at Hickson & Welch Ltd* [HSE 1992]; *The chemical release and fire at the Associated Octel Company* [HSE 1994a]; *The Southall Rail Accident Inquiry Report* [HSC 2000]). For example, the report into the explosion and fire at the Texaco Refinery at Milford Haven (HSE 1994b: 1) noted that, 'the direct cause [...] was a combination of failures in *management*, equipment and control systems during the plant upset (emphasis added)'.

The second edition of HSG65 (HSE 1997) has a greater emphasis on leadership, but with the objective of successfully *managing* health and safety. According to the guidelines, management needs included leadership skills. Managers were now to lead. While this guidance was still aimed at directors, it now extended to senior managers with health and safety responsibilities. The scope of those charged with managing health and safety had broadened. The publication of the Turnbull Report (ICAEW 1999) encouraged the HSE to emphasize the importance of visible leadership of health and safety by senior managers, with companies ideally appointing a champion on their board of directors (Eves 2014). Taken together, these reports promoted the idea of safety leadership amongst the senior managers of organizations, a point emphasized in the Cullen Report (2001) investigating the Ladbroke Grove rail disaster and in the HSE publication *Leadership for the major hazard industries* (HSE 2004). This aimed to 'refresh your knowledge of effective health and safety leadership', based on various models of leader engagement with followers. However, the notion of 'safety leadership' had not yet become mainstream. The HSC's publication, *A strategy for workplace health and safety in Great Britain to 2010 and beyond* (HSC 2004a: 8) contained no mention of safety leadership but emphasized that 'appropriate health and safety management is an integral part of effective business management and as such, is an enabler and not a hindrance'. Moreover, the British Standard (BS8800; BSI 2004) referred to health and safety leadership by top management, not in the main text but rather in Annex C.1 General (p. 23). In the early 2000s, then, we might be excused for asking whether safety is being managed by directors and other senior managers, or being led by someone else? And if so, by whom?

'*Be part of the solution*' (HSE 2009) unequivocally stated that 'health and safety leadership is all about accountability' (p. 9) and that 'following the example of leadership at the board level, leadership must also permeate throughout the management and supervisory levels and the workforce' (p. 9). This reaffirmed the guidance in BS18004 (BSI 2008a) where the top management team should 'provide clear and visible leadership on OH&S' (p 20), and in OHSAS 18002 (BSI 2008b), where 'top management should demonstrate the leadership and commitment necessary for the OH&S management system to be successful and to achieve improved OH&S performance' (p. 11). From this point, according to the HSE, everyone is potentially a safety leader.

This may actually be truer to the spirit of the *Health and Safety at Work etc. Act* (1974: section 7), namely 'It shall be the duty of every employee while at work to take reasonable care for the health and safety of himself and of other persons who may be affected by his acts or omissions at work'. By emphasizing

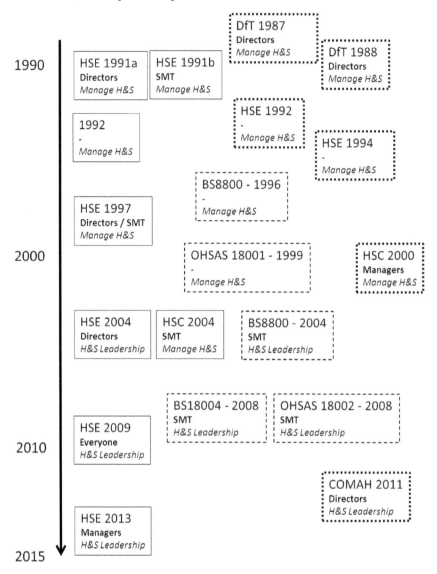

*Figure 6.1* Timeline of emergence of term 'Safety Leadership' in selected guidelines for practice (solid borders), standards (dashed borders) and accident investigation reports (dotted borders)

accountability, which transcends the traditional scope of a manager's activities (see Hales 1999), safety is now the responsibility of a leader. Subsequent guidance, including the third edition of HSG65 (HSE 2013), reinforces this. The COMAH report (COMAH 2011) into the earlier explosion and fire at the Buncefield oil storage depot in 2005 focused mainly on technical aspects of safety and safety management. It responded to this new safety leadership agenda by merely observing, in a non-specific recommendation that:

> clear and positive process safety leadership is at the core of a major hazard business and is vital to ensure that risks are effectively managed. It requires board-level involvement and competence. Board-level visibility and promotion of process safety leadership is also essential to set a positive safety culture throughout an organization.
>
> (p. 29)

This unhelpfully introduces yet another form of leadership – 'process safety leadership'.

This brief history brings out two problems. First, if safety is the responsibility of a leader, and everybody is potentially a leader, then there is a real risk that nobody will show leadership, and simply assume that somebody else will take responsibility. Consequently, safety performance may deteriorate. Second, by emphasizing 'safety leadership', safety becomes a separate stand-alone activity carried out by specialists. This is exemplified in the appointment of 'safety champions', whether in the workplace or in the boardroom, as envisaged in the HSE's guide for directors which 'recommends that every board should appoint one of their number to be a "health and safety" director' (HSE 2002). Such an approach inevitably divorces safety from everyday workplace activity and works against the development of a safety culture and improved safety performance.

### Practices of safety leaders: evidence from the literature

According to the HSE-sponsored literature review of effective leadership behaviours for safety conducted by Lekka and Healey (2012) much of the current safety leadership research is focussed on *transactional* and *transformational* leadership.

Transactional leadership is based on non-individualized hierarchical relationships and comprises three dimensions (constructive leadership, corrective leadership and laissez-faire leadership; Zohar 2002a). Constructive leadership offers material rewards (e.g. increased salary, promotion, job security) contingent upon satisfactory performance. This requires clear communication between leader and follower. Some understanding of the needs and abilities of individuals is required in order to offer rewards that match their motivations. Corrective leadership (or active management by exception) monitors individual performance against standards, detecting errors and correcting them. Laissez-faire leadership (passive management by exception) disowns all leadership responsibility and only engages with subordinates in an emergency.

Based on an ABC (Antecedents–Behaviours–Consequences) model, studies by Zohar and colleagues (Zohar 2002b; Luria et al. 2008; Zohar and Luria 2003) showed that supervisors were providing workers with verbal and non-verbal feedback (both positive and negative) on their performance of safe working practices. This feedback drew on prior training and goal setting (the antecedents to the desired behaviours). Transactional safety leadership practices of supervisors therefore included:

- establishing and communicating appropriate safety goals;
- monitoring performance towards these goals; and
- rewarding (through giving feedback) behaviours that sustain or improve safety performance.

These three items for transactional safety leadership practices were confirmed by Kapp (2012b).

Transformational leadership may be defined as leader behaviours that transform or inspire followers to perform beyond expectations, while transcending self-interest for the good of the organization (Avolio et al. 2009: 423). Transformational leadership comprises four leader behaviours (Bass 1985): idealized influence; inspirational motivation; intellectual stimulation; and individualized consideration. It is characterized by value-based and personalized interaction, which results in better exchange quality and greater concern for welfare (Zohar 2002b). Studies focusing solely on transformational safety leadership come from the work of two groups. A Canadian team (Barling et al. 2002; Kelloway et al. 2006; Mullen and Kelloway 2009) has studied safety leadership mainly in service sector settings, while a UK-based group (Conchie et al. 2012; Conchie and Donald 2009; Conchie et al. 2013) studied the role of trust in the relationship between supervisors and workers in high-hazard settings. The key elements of transformational safety leadership (Kelloway et al. 2006) include:

- expressing satisfaction when jobs are performed safely;
- rewarding achievement of safety targets;
- continuous encouragement for safe working;
- maintaining a safe working environment;
- suggesting new ways of working more safely;
- encouraging employees to openly discuss safety at work;
- talking about personal value and beliefs in the importance of safety;
- behaving in a way that demonstrates commitment to safety;
- spending time to demonstrate how to work safely; and
- listening to safety concerns.

A number of research studies that investigate the co-occurrence of both transformational and transactional safety leadership (Zohar 2002a; Dahl and Olsen 2013; Clarke and Ward 2006) suggest that safety performance (i.e. the reduction of injuries) and safety compliance is positively related to transactional safety leadership practices. Safety behaviours (i.e. safety participation) on the other hand are encouraged by transformational safety leadership (e.g. concern and motivation).

Table 6.1 identifies eight leadership practices that may affect safety outcomes and which have been reported by key authors in *Professional Safety*, the *American Society of Safety Engineers* (ASSE) journal. According to these articles, safety leadership is fraught with difficulty. This is partly a result of the commonly perceived tension between safety and productivity (Carillo 2005), which creates ambiguity, and partly because safety leaders are often squeezed between senior leaders and operational managers (Forck 2012).

A number of policy research reports, mainly published by the HSE, have also looked at safety leadership and identified leadership practices that, either directly or indirectly, stimulate others, mainly front-line workers, to deliver positive safety outcomes. Indirect effects were achieved by modelling appropriate safety behaviours (Poxon et al. 2007), or setting agendas, or safety goals and targets, for others to follow (e.g. King et al. 2010). More direct effects of leaders occur through immediate challenge (Cummings 2006) or engagement with the workforce (Healey and Sugden 2012; Cummings 2006; Busby and Collins 2009), demonstrating the value of the employee to the leader (Poxon et al. 2007), ensuring effective two-way safety communication (Fleming 1999), and motivating employees (King et al. 2010). Direct effects also occur through empowering employees to problem solve and to make decisions (Fuller and Vassie 2005). Finally, direct effects can occur through developing skills, especially in independent and inter-dependent working (Poxon et al. 2007), and knowledge of safety practices (Lekka and Healey 2012). These are summarized in Table 6.2 and categorized according to the four dimensions of transformational and transactional leadership.

Many of these practices are echoed in the literature reviews found in other policy reports. Lekka and Healey (2012), for example, note that studies show the importance of leader support for safety and safety communication between management and workforce. Active involvement in safety, and enforcement of safety, promotes perceptions of a positive safety climate and fosters employee accountability and responsibility. Gadd and Collins (2002) concur. They observed that management commitment to safety reduced under-reporting of incidents and promoted a positive safety culture, but often without indicating precisely how this was achieved. O'Dea and Flin (2003) developed a descriptive model that shows how safety leadership and the required actions differ according to the leader's level in the organizational hierarchy.

- Senior managers demonstrate safety leadership through:
    - positive attitudes to safety by committing to safety policies and procedures;
    - ensuring safety is integral to competitiveness and profitability and safety;
    - assuring safety compliance; and
    - committing to developing trusting relationships with subordinates.

Table 6.1 Leadership practices identified in articles published by different authors in the practitioner journal *Professional Safety*

| Leadership practices | Drennan and Richey (2012) | Krause and Weekley (2005) | Kapp (2012b) | Geller (2000, 2008) | Mathis (2013) | J.H. Williams (2002) | Forck (2012) | Petersen (2004) |
|---|---|---|---|---|---|---|---|---|
| Set goals/define roles | x | x | | x | x | x | | |
| Monitor performance | x | x | | | x | | | x |
| Educate/train | x | x | | x | | | | x |
| Role model | x | | x | | x | | | x |
| Communicate (share information, seek ideas and opinions, listen) | | x | | x | | x | | x |
| Involve others/participation/collaboration | | | | | x | | x | |
| Show care/concern/interest | | | | | | x | | x |
| Reward/give feedback/recognize good work | x | x | | | x | | | x |

- Middle managers (or site managers, typically in construction or oil platforms, which have been the focus of much research) show safety leadership through:
  - demonstrating commitment to safety by interpreting and implementing safety policies positively;
  - prioritizing safety in work planning and scheduling;
  - being actively involved in safety by being visible in taking responsibility;
  - communicating openly; and
  - showing concern and appreciation for employees.
- Safety leadership is demonstrated by supervisors and team leaders:
  - by support (giving open and fair feedback);
  - by involvement (in safety training, inspections and meetings); and
  - by being participative (encouraging teamwork and building trusting relationships).

Recent HSE guidelines (e.g. HSE 2012) differentiate the leader and the manager and their respective responsibilities for safety. Similar distinctions are made in *Managing for health and safety* (HSG65) (HSE 2013). This suggests that safety leader roles can be differentiated from safety manager roles, and that observed safety practices can be similarly disaggregated. O'Dea and Flin's (2003) review noted a similar differentiation of safety roles and practices across the hierarchy of an organization. A comparison of the practices of safety leaders with the 11 features that characterize managerial work (Hales 1999) reveals considerable overlap between the two lists, implying that what is often described as safety leadership practice is actually management practice.

Rather than debating whether these practices are leadership or management, it may be more helpful to collate them and capture the underlying safety objectives. Building on earlier suggestions identified in the ACSNI report (HSE 1993), developed independently by Wu et al. (2008), and reiterated in *A review of the literature on effective leadership behaviours for safety* (Lekka and Healey 2012), we propose three cross-level categories of safety leadership behaviours (safety controlling, safety caring and safety coaching). Using these three categories, we can assemble the empirical findings from research, policy and practitioner literatures to develop a generic list of practices, covering different themes, for use in diagnosing safety enactment in organizations (Table 6.3).

### Practices of safety leaders in service sector organizations: empirical evidence

The practitioner and policy literatures and research reports identify a variety of practices enacted by those designated as safety leaders, mainly supervisors and team leaders. Other guidelines (e.g. HSE 2013) define practices required to comply with legislation. These include writing safety policies, assessing risks, controlling and monitoring risks, training, instructing and supervising.

Table 6.2 Leadership practices identified in the empirically-based policy reports as they map onto the dimensions of transformational–transactional leadership

| Authors | Transformational leadership dimensions | | | | Transactional leadership |
| | Idealized influence | Inspirational motivation | Intellectual stimulation | Individual consideration | Constructive leadership |
| --- | --- | --- | --- | --- | --- |
| Healey and Sugden (2012) | Consistent implementation Role modelling | Clarity Develop a safe environment Team working Prioritize safety | Involve others Listen Empower others | Train | Reward Set goals Monitor Give feedback |
| Conchie and Moon (2010) | | Communicate | Voice Communicate | Coach Show concern | |
| King et al. (2010) | Set an example Be visible | Motivate others Prioritize safety Communicate | Seek feedback Consult Communicate | Care | Set goals Monitor |
| Busby and Collins (2009) | Consistency | Prioritize safety | Engage others | Develop working relations | Provide resources |
| Poxon et al. (2007) | Model behaviours | Define issues Share agenda | Empower others Communicate | Value others Develop others | Set goals |
| Cummings (2006) | | | Encourage commitment | | Challenge behaviours |
| Fuller and Vassie (2005) | Be responsible | Motivate others | Communicate Engage others Problem solving Decision-making Involve others Communicate | Train | |
| Brazier et al. (2004) | | | | | Plan Monitor performance |
| Fleming (1999) | Be visible | Prioritize safety | Participative decision-making | Value others | |

*Table 6.3* Safety leadership practices synthesized from different sources based on a Coaching–Caring–Controlling model of safety leadership

| | Source of safety leader practices | | | Summary generic practices |
|---|---|---|---|---|
| | *Academic* | *Policy* | *Practice* | |
| Safety coaching | Demonstrate commitment Prioritize safety Encourage open discussions Talk about values and beliefs Provide direction Problem solve | Role model Set an example Be visible Prioritize safety Motivate others Involve others Empower others Coach | Role model Educate Train Involve others/ participation | Role model Prioritize safety Involve others Empower others Train Coach Be visible |
| Safety caring | Provide support Express satisfaction Listen to safety concerns Demonstrate how to work safely Support Care Show concern Maintain a safe working environment | Share agenda Communicate Develop a safe environment Listen Care Consult Develop working relationships/ team work Value others Develop others | Show care/concern Communicate (share information/seek ideas/listen) | Communicate Listen Show concern Care Support Create and maintain a safe working environment Value others Develop others |
| Safety controlling | Monitor Reward | Assure compliance Reward Set goals Monitor performance | Set goals/define roles Monitor performance Reward/give feedback | Set goals Monitor performance Reward |

These have a more managerial orientation, reflecting the administrative tasks embraced by the acronym POSDCORB (Planning, Organizing, Staffing, Directing, Coordinating, Reporting and Budgeting), proposed by Gulick (1936) to distinguish management from other roles.

Our study (Pilbeam et al. 2015) identified a range of practices deployed by managers, supervisors and front-line workers in service sector organizations, where there has previously been little research. Previous work on safety leadership has concentrated on organizations in the energy and manufacturing sectors (e.g. Clarke and Ward 2006; Dahl and Olsen 2013; Zohar and Luria 2003) and neglected those sectors where most people actually work in contemporary developed societies.

From 143 interviews, we identified 41 different leadership practices. Figure 6.2 shows the 24 practices reported by more than 10 per cent of the interviewees. The most common, reported by more than one third of the

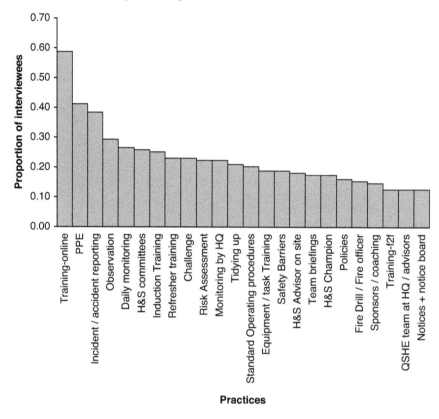

*Figure 6.2* Proportion of interviewees identifying different safe work practices in three different sectors. (Pilbeam et al. 2015)

interviewees, were on-line training, wearing PPE, reporting accidents and near misses, and the observation of work practices.

The practices that ensured the delivery of safety in these organizations may be differentiated by the role of the interviewee in the organizational hierarchy (Figure 6.3). Some practices were mentioned by people in all three roles: training, protecting, investigating and reporting, and observing. However, there also seem to be some practices that are more or less exclusive to people in each of the roles. Front-line workers said they were told what to do through briefings and to focus on good housekeeping and respect safety barriers designed for their protection. Supervisors focussed on ensuring front-line workers were working safely by making sure their training was up to date, challenging poor practices and providing support through coaching and mentoring where necessary. They also played some part in the daily inspection of the local work environment, and were aware of, even if not necessarily involved in, the regular tracking books submitted to headquarters. Managers referred to a greater range of practices including: setting procedures; developing policies; dealing

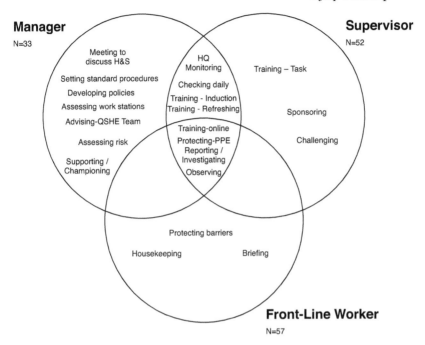

*Figure 6.3* Safety related practices reported by 20 per cent or more of interviewees in each role. (Pilbeam et al. 2015)

with audits; seeking advice from the QSHE team; discussing health and safety at regular meetings; and assessing risks.

## Alternative modes of safety leadership

The current mode of safety leadership, building on the *Health and Safety at Work etc. Act*, 1974, and subsequent guidelines from the HSE has a 'command and control' ethos. The directions of the few are supposed to be followed uncritically and systematically by the many. Board level and senior managers clearly make an important contribution to the development of a safety culture, the establishment of a safe working environment and the implementation of safe working practices. In practice however, this contribution may be less apparent. We found that about 10 per cent of our total sample were unclear about 'who sets the safety policy and goals' in their organizations. In some retail organizations, up to one third of interviewees did not know where the policies and goals originated. In the office environment, interviewees, 'imagine[d] that the QSHE team do that'. 'The HSE team I would have thought, I don't know of anybody specifically'. This lack of certainty contrasted with the much stronger conviction that individually they had no involvement, 'no, no, no'. It was all somewhat vague: 'we usually get an email when there's something new'. If interviewees knew where the policies came from, they 'imagine[d] most of it comes from corporate' or simply from 'someone up above'. When they did not know, or thought it came from their manager by a process of 'filtering down', they were 'just told to follow procedure and that's what I do'. Policies from head office were followed passively, with only limited scope for local adaptation: 'we follow the company guidelines. No matter how big or small your store is you don't get the opportunity to flex or opt in or opt out'. 'Each store will have the same Health and Safety policy' or 'we just follow the [organization's] overarching policy'.

In contrast to the retail and office sectors, where at least 60 per cent of interviewees felt that safety policies came from head office, in the warehouse, less than 40 per cent thought policies came from head office. They identified greater input from local health and safety advisers, which was absent elsewhere. The greater awareness of the health and safety adviser may reflect their full-time paid role, but also the potential to develop policy locally within the context of generic company-wide safety policies. A health and safety adviser at one warehouse noted that that 'the policy for this site may be slightly different to what you'll see over next door, because they've got different risks'.

We suggest that there may be forms of safety leadership other than command and control. These may be more responsive to different aspects of safety legislation or the nature of safety itself. First, safety legislation requires everyone to take care of themselves and their fellow worker. This plays to the more recent notions of plural forms of leadership, particularly distributed leadership. Second, safety may be enacted 'in the moment', as noted in the two previous chapters from colleagues at Loughborough, suggesting that a phenomenological approach may help our understanding of safety leadership. Third, safety, because of its systemic nature, may be defined as an adaptive challenge requiring adaptive leadership. We shall consider each of these possibilities in turn, drawing out the required leadership characteristics and the implications for training and development.

### Plural leadership

Recent reviews of leadership (e.g. Thorpe et al. 2011) have considered approaches that reach beyond the earlier unitary views of leaders as individuals – the model of the heroic leader. By recognizing that leadership skills and responsibilities can be dispersed or shared throughout an organization, these perspectives focus on the process of leadership rather than on the person as leader (Gordon 2002). Denis et al. (2012) describe these leadership forms as 'plural leadership'. They identify three strands in this literature that make sense of its otherwise confusing terminology, where labels like 'shared', 'distributed', 'collective', 'collaborative', 'relational' or 'post-heroic' are often used loosely and interchangeably.

The first strand of 'plural leadership' considers *mutual* or *shared* leadership within a group or team, where members collectively lead each other. This participatory approach is encouraged by transformational leadership and is consistent with earlier studies of the emergence of leadership in groups (e.g. Bales and Slater 1955). These studies noted the need for individuals to play different and complementary roles, embracing 'task functions' and 'expressive functions' in group leadership. This model demands that individuals are motivated to share leadership responsibilities and that opportunism or 'free riding' is discouraged. Shared leadership has a distinct application in team-based organizations, which are common in low-hazard environments. It involves sustaining a shared appreciation for the importance of safety and giving equal value to those responsible for this seemingly 'lesser' activity alongside those who lead more creative and exciting tasks (Table 6.4). Team working skills, notably valuing others and taking responsibility, are especially important for delivering safety leadership in this mode.

A second strand of 'plural leadership' explores the circumstances where a small number of individuals pool their leadership capacities to co-lead others. Here the co-leaders play roles that are specialized (i.e. each operating in particular areas of expertise), differentiated (i.e. avoiding overlap) and complementary (i.e. cover all the required areas of intervention). Gronn (2002) suggested that they conjointly exert leadership, having a collectively agreed and common purpose, characterized by reciprocal influence. Currie and Lockett (2011) also include concertive action, where skills and expertise are pooled, permitting individuals to work together closely within a framework of shared understanding, often developed implicitly. It can be challenging to achieve both *conjoint agency* and *concertive action*, particularly in relation to safety where the co-leaders may disagree over its importance and the best way to achieve it. Skills of negotiation, listening and communication are required (Table 6.4).

A final strand of 'plural leadership' is *distributed leadership* (Fitzsimmons et al. 2011; Spillane 2006), where leadership roles are dispersed or spread across organizational levels over time, so that many people can take on leadership roles at appropriate moments. Distributed leadership, insofar as it is seen to

*Table 6.4* Key challenges and skill requirements of safety leadership in 'plural' mode

| Type | Key safety leadership challenges | Skills |
|---|---|---|
| Shared | • Maintaining safety as a priority<br>• Ensuring technical skills and knowledge of safety requirements are current | • Valuing others<br>• Taking responsibility<br>• Communication |
| Co-leadership | • Reaching agreement on priority of safety<br>• Sharing a clear view of the way of working safely | • Negotiation<br>• Listening<br>• Adaptability |
| Distributed | • Awareness of safety competencies throughout the workforce<br>• Promoting the freedom to challenge<br>• Maintaining ownership of leadership responsibilities | • Deferring to expertise<br>• Appreciating others<br>• Challenging others |

be democratic, encourages collective capacity-building, and increases efficiency and effectiveness by making better use of expertise (Mayrowetz 2008). Obviously this model of leadership demands that everyone is both aware of and respects the different skills and competencies found throughout the organization. It also requires a reduction in power differentials between individuals and groups so that individuals are able freely to challenge others and to assume leadership responsibilities as required (Table 6.4).

### Leadership 'in the moment'

In her book *Rethinking Leadership*, Ladkin (2010) introduced a model of the 'leadership moment', which 'conceptualizes the interactive and context-dependent nature of leadership' (p. 27), identifying four elements that interact in the experience of leadership. These elements are leader, follower, context and purpose. Leaders and followers must relate to each other within a particular context as together they pursue a common purpose. These elements interact dynamically so that the followers' perceptions of context will affect their interpretation of the leader's pronouncements and the leader's behaviour will be affected by the followers'. Their combined actions show how the purpose is being understood and lived out.

While a reduction in injuries, the creation of a safe working environment or improvements in workforce safety behaviours might constitute the apparent purposes of safety leadership, there has been little research on how these are actually achieved, and how this might differ with context, the degree of follower involvement or the leader's ambition. Our colleagues at Loughborough studied this through direct observation, as they describe in the two previous chapters. This approach is less suited to the workplaces we studied so we asked employees in stores of two retail chains to keep audio diaries identifying when safety became salient to them during their working day. Some of their responses demonstrated 'leadership moments': for example, when employees chose to relocate or reposition items of stock that were creating potential trip or fall hazards in the stockroom, or when they helped others lift heavy items. In both of these circumstances, possible injuries were prevented and safer working assured.

From this perspective, 'safety leadership' will remain elusive because it depends on the circumstances in which it is enacted: each 'moment' contributes a small piece to our understanding of the whole. Nevertheless, as Ladkin (2010) suggests, this approach may help us to better engage with 'leadership' in a safety context by clarifying what we need to know about the elements of the 'leadership moment', their interrelationship and subsequent contribution to safety. Are we interested in better understanding how to lead the members of our team to work safely? Are we trying to find out why a particular set of leadership practices designed to achieve safety outcomes worked in one setting but were less effective in another? Are we interested in understanding how the same safety outcomes may be achieved in so many different ways?

## Adaptive leadership

Organizations typically develop safety policies and standard operating procedures anticipating that these will provide respectively a safe working environment and safe working practices. Accidents, injuries and near misses still happen. Clearly, these 'technical solutions', although necessary and beneficial, are insufficient on their own to achieve safety completely. Safety is a problem that arises from the multiple interactions between both human and technological components of a system. Such systemic problems create adaptive challenges, which call for adaptive leadership (Heifetz and Laurie 1997). Systemic problems often demand changes to organizational values and beliefs. Prioritizing safety is one such value that often runs counter to traditional organizational cultures which emphasize productivity, standardization, innovation and creativity or personal development (Denison and Spreitzer 1991). Solutions to the problems of embedding safety (like other systemic problems) lie throughout the organization and are not merely the responsibility of the director or health and safety manager.

Heifetz and Laurie (1997) identify six principles for leading adaptive work:

- '*Getting on the balcony*'. Stepping back from the day-to-day detail to assess the bigger picture.
- *Identifying the adaptive challenge*. Fully understanding the nature of the existing problem.
- *Regulating distress*. Inspiring change and allowing employees to debate issues and clarify assumptions before providing direction so that people are not disabled but challenged.
- *Maintain disciplined attention*. Grapple with the issues, digging deeper into the conflicts emerging from different perspectives on the problem and encourage collective problem solving.
- *Give the work back to employees*. Supporting and empowering individuals to take both risks and responsibility to find the solution to the problem.
- *Protect voices of leadership from below*. Don't silence the whistle-blower or the deviant, rather take time to explore why they are doing what they do.

Each of these can be applied to the practice of safety leadership, enabling responses to specific safety-related questions and particular safety challenges (Table 6.5). The table also suggests a number of abilities that adaptive safety leaders would need to display to be effective.

## Conclusion

Delivering safety in organizations is prescribed by government regulations and informed by guidelines promoted by the *Health and Safety Executive*. These largely dictate the diversity of practices deployed by safety leaders in organizations. In part, this diversity of leadership practice is a result of the

*Table 6.5* Adaptive leadership and its application to safety leadership

| Principles of adaptive work | What are the challenges for a 'safety leader'? | Suggested abilities/ characteristics of safety leadership | Key safety question/ issue to be addressed |
|---|---|---|---|
| Get on the balcony | Understanding the interactions of the system | Ability to take an holistic view, to integrate | What contributes to safety? |
| Identify the adaptive challenge | Clarifying the safety problem/issue | Ability to interrogate multiple sources of data | What is the cause of the particular safety problem? |
| Regulate distress | Managing discomfort/disbelief/ resentment of others | Ability to live with uncertainty and frustration | How do I motivate others? |
| Maintain disciplined attention | Surfacing differences of opinion/priority about safety | Ability to draw complementary skills together, to facilitate dialogue | What conflicts with safety? |
| Give the work back to people | Empowering others to solve safety problems ... take ownership of safety | Ability to remove distractions and focus on key issues | What skills are available to solve this safety problem? Who can help? |
| Protect voices of leadership from below | Listening to the views of others | Ability to consider alternative perspectives | What's really going on here? |

historical overlaying of leadership onto those managerial practices formerly used to ensure safe working. Rather than emphasizing either safety leadership or safety management, which inevitably separates safety from mainstream organizational activity, we suggest that safety may be more effectively delivered in organizations by collating the existing (and future) practices according to three evident safety objectives – Caring, Coaching and Controlling. Not to be confused with the 4Cs for promoting a safety culture (HSE 1997), namely control, cooperation, communication and competence, the 3Cs of safety leadership behaviour could provide a useful diagnostic tool for safety enactment in organizations.

Hitherto, safety leadership has followed individualistic transactional or transformational models of leadership. While these may be relevant to more hierarchical organizations, or where there is a heavy emphasis on supervisory control, they may be much less applicable in networked organizations or where there is a dominance of professional workers. Both of these are characteristics of the service sector. In these circumstances, recent conceptualizations of 'plural' leadership may be more appropriate for delivering safety. Simple technical solutions may not resolve the organizational challenges created by the increasing turbulence of the organizational environment. Assuring the effective delivery of organizational safety is one such challenge. Harnessing the principles of adaptive leadership to safety problems may ensure a safer working environment for all.

## Key points

- The concept of 'safety leader' is confused and makes little use of the wider understanding of leadership in organizations that has emerged. In particular, the focus on leaders has emphasized individual characteristics at the expense of leadership, which underlines the importance of social relationships and interactions.
- Leadership in relation to health and safety has different properties from leadership in other areas of management. This has marginalized health and safety within organizations as a specialist responsibility rather than something all line managers need to be engaged with.
- The fashion for presenting health and safety as everyone's responsibility means, in practice, that it easily becomes no-one's responsibility.
- Much current thinking on health and safety leadership is based on command and control models that are less relevant and less effective in contemporary networked organizations. Plural models of leadership that rely less on the enforcement of rules and policies and more on employee engagement and empowerment may be a better basis for achieving health and safety objectives.

# Conclusion: stepping up to the challenge

*Shelley Frost and Robert Dingwall*

Public perceptions of health and safety seem to have become more polarized over the last 20 years. The profession could explain this by pointing to the role of successive governments in dismantling the social contract that created the context in which the *Health and Safety at Work etc. Act* could be passed. This has been magnified by the UK media's treatment of 'elf and safety' as a subject for ridicule. The social, cultural and technological legitimacy of the profession has been undermined by the priority given to commercial interest in difficult economic times. The profession should, though, be prepared to ask what part its own members might have played in health and safety's slide from core social value to compliance policing. Do at least some of the public and political problems with health and safety have a basis in the profession's own behaviour? Is the profession itself doing enough to ensure that all organizations benefit from well managed occupational safety and health?

Why do organizations treat health and safety professionals differently from other business critical specialists? It is not just a media story, or 'regulatory myth', that the health and safety person is a 'jobsworth' or a 'work prevention officer'. This attitude does still exist in some organizations. Health and safety is not given the same status as other core functions. It is not uncommon to find people employed in health and safety roles who lack the requisite skills, knowledge and experience to deliver this function effectively. Organizations feel comfortable doing this because they see health and safety simply as a compliance issue: if you only want to answer the question 'are we compliant?' you only need someone who can go down a check list.

Lagging metrics, particularly deaths and lost time, are certainly the prevalent approach to performance measurement for health and safety. In a pressured environment where everyone needs to 'get it done', there seems to be very little room for health and safety to do more than tick compliance boxes. However, nobody goes into health and safety expecting to spend their time filling out forms, digesting regulations and policing rule breakers. Every health and safety professional experiences the frustration of knowing that a bit of proactive activity – a little more planning and forethought – could avoid troubles and delays. This is good business management and 'best practice'.

What can the profession do to make health and safety a core business value again? Fundamentally, health and safety is about people. Ask any health and safety professional about the beginnings of their personal journey into the profession. What you will hear is a passion for protecting life and eliminating harm that has become entangled with the administration of a set of laws and regulations. This passion is the key to the profession's potential contribution to a rapidly evolving world of work. Faster, higher, stronger is not just the Olympic motto but a mantra repeated in boardrooms around the globe. Everything has to be 'agile', everything has to be 'lean' and everything has to be right first time 24/7/365. Successful twenty-first century businesses need to understand, and extract full value, from every element of their business, including their workforce.

At the same time, the general public are developing increasing expectations about the behaviour of organizations that they are willing to engage with. Those expectations – the social licence to operate, in the words of Chapter 1 – temper the routine actions and strategic decisions of every organization. One powerful example is the way environmental change movements have fostered societal expectations that all organizations will seek to minimize their impacts on the environment. This has become a significant area of competitive advantage. Organizations derive both reputational and economic benefits from counting the 'carbon cost' of doing business and showing they are working to become 'carbon neutral'. They go to great lengths to demonstrate how they are improving peoples' lives while doing business – corporate social responsibility has become an industry worth billions of pounds, dollars or euros, driven by the expectations of the society which organizations inhabit.

It should not be surprising that these movements have also changed the way governments and politicians interact with business. As organizations become more responsive to societal expectations, it is increasingly assumed that they will take responsibility for 'doing the right thing'. There is a shift away from providing detailed central prescription or guidance on the ways to achieve this, although there may be stricter legal or reputational penalties for those who do not deliver the desired outcomes. While most businesses welcome the increased flexibility in deciding how to achieve the expected goals, this does require them to accept a corresponding responsibility for the ultimate social impact of their commercial decisions.

The field of health and safety is no exception. Regulatory changes, as we saw in Chapter 2, have placed increased emphasis on the responsibility of organizations, and individuals, to deliver the outcome of a safe workplace rather than micro-managing the means by which this is achieved. Organizations are expected to translate a 'safe workplace' into their own operational targets and to recognize the wider benefits of setting and meeting standards beyond mere compliance. A healthy and motivated workforce demonstrates the organization's values, contributes to its reputational capital and, ultimately, contributes to the bottom line and sustainability of the business.

Some organizations have already embraced the 'zero harm' or 'zero tolerance' approach to health and safety. They have recognized the challenges coming from their societal environment. Their safety and health professionals are playing a key role in promoting the engagement of all employees towards these goals. These organizations want their professionals to offer leadership: to be able to define what 'good' looks like, to monitor their progress, to recognize when they have achieved a goal – and to sustain that achievement. In order to do these, the organizations need to be able to identify competent professionals and to understand what creative skills, knowledge and experience they are bringing with them. These professionals, as Chapter 3 discussed, are the links between an organization and a wider community of shared evidence, learning and understandings of good practice.

Every organization is different so it is important that professionals do not assume that there is only one answer to any given question. They need a portfolio of competencies to be able to devise the right solution within a specific context, and to be able to continue searching for incremental improvements. Those improvements may be driven by performance measurements and benchmarking within and between organizations and their partners. Equally, however, as we saw in Chapters 4 and 5, they may also be driven by careful attention to the improvisations of the workforce as they search for ways to manage the problems that they encounter. Successful professionals as leaders are good listeners as well as good advocates. The behaviours of leaders determine business performance, and ultimately results. Contemporary leadership, as Chapter 6 showed, is about playing a constructive and flexible role in a team rather than an exercise in personal ego.

It is against this background that IOSH has been reviewing its own contribution to the continuing development of a safety and health profession that can meet the needs of the twenty-first century. IOSH has itself sought to model good practice in leadership through wide consultation and engagement with its own members and other stakeholders. The result is a free to access and free to use 'competency' scheme which will:

- provide clarity and consistency about the elements of skills, knowledge and experience required for people taking on OSH responsibilities;
- specify the need for both technical and business, communication and strategic competence in order to carry out OSH duties effectively;
- provide a mechanism at an international level for characterising what 'good' looks like to enable benchmarking amongst the profession, across organizations and between countries;
- provide a consistent professional development pathway – enabling users to identify both current competence and development needs to meet their ambitions;
- identify the potential routes for professionals to fill development gaps, such as training programmes, publications, best practice case studies, reference material, and so on;

- provide a common language on occupational health and safety, recognized internationally and across business sectors;
- contribute to establishing common metrics in measuring OSH to ensure consistency and parity across organizations.

This will be the first comprehensive package to enable people with OSH responsibilities, and the organizations that they serve, to assess, plan and develop competency needs as the basis of a personal or organizational development programme. The scheme recognizes the increasing use of 'team-based' structures and allows organizations of any size to assess and identify development needs for the individual, team and organization as a whole. If adopted in full, the scheme will provide necessary evidence on the 'true value' of health and safety to the organization and ensure professionals possess the right expertise (including engagement skills) to bring about change. The results will be seen through their impacts on the organization's reputation, resilience and, ultimately, performance.

The competency scheme will also offer a means of linking the increasingly complex networks that are involved in many manufacturing, construction and service industries. The decline of large, vertically integrated, organizations represents a challenge to the achievement of consistent health and safety standards across all the collaborators whose work is involved in delivering a particular product or service. It is intended that the scheme should provide an opportunity for the collaborators to determine a standard set of performance metrics and to validate each other's contribution by reference to common benchmarks for the knowledge and competence of the professionals employed by each party.

The world of work is constantly changing and developing. The novel or fringe idea of ten years ago becomes the common business practice of today. CSR, sustainability, well-being, business risk management and reputational assurance are now all part of the landscape that drives core organizational values. Safety and health can be a thread that runs through all of these, in the way that it recognizes the workforce as a crucial organizational asset. An organization that is indifferent to the human capital represented by its employees is ultimately unsustainable. The profession needs to engage organizational leaders and stakeholders in understanding how health and safety is fundamental to the requirements of strategic visions under the conditions of the twenty-first century.

The standards being developed in the competency scheme will challenge professionals to develop the social science skills of management and engagement to the same level as the scientific, engineering and biomedical skills that they have traditionally used. As the research presented in this book has documented, the modern health and safety professional needs to have a much broader and more structured understanding of society, organizations and people than that gained solely through experience. They need explicit tools and language to make the case to other team members for investment in their work. Practical or clinical experience alone will be insufficient to develop a coherent, well-argued and evidence-based justification for the contribution of health and safety to the overall performance of the organization. The profession, for example, needs to develop better leading indicators for health and safety performance, rather than relying on the lagging indicators of accidents or deaths. Closer engagement with the sources of explicit knowledge, as discussed in Chapter 3, should assist with this. As Chapters 4 and 5 showed, there are also important challenges in developing a range of implementation and communication skills to promote change at all levels of an organization or network, in addition to the technical skills of identifying which areas need to change. Following the work in Chapter 6, we can also see how adaptive leadership will be required, which is not restricted to a handful of people at the top or specified health and safety champions but which enlists the whole workforce in a collective project to create a safe and healthy workplace for all. Professionals need to understand how and when to empower and enable their colleagues rather than commanding them.

The competency scheme will be 'owned' by everyone and 'belongs' to all who can benefit from it – it encourages the exchange of ideas, introduction of new ideas and a continuing drive to adapt to current needs. It is not designed to favour large organizations: in fact, it should be most transformational for SMEs. The scheme will enable them to be clear on how they can support their workers and ensure their staff have the competencies to achieve their health and safety objectives. By focussing on individuals, the framework can be adapted to the specific context of any particular SME and can be shaped by its unique culture and inter-relationships.

More broadly, the scheme will help the profession to speak to wider publics with a single, informed and authoritative voice about the positive role of health and safety (with proportionate risk-based messaging). Stakeholders at all levels will naturally be engaged in the collaborative and flexible development of health and safety policy with opportunities to influence the agenda and outcomes. In effect, IOSH is demonstrating the leadership process that it will be encouraging its members to adopt. By showing more adaptive leadership, IOSH hopes to address some of the issues about public confidence identified in Chapters 1 and 2 of this book and to offer a more positive message about the profession's contributions and achievements. This will support a better informed, less prescriptive, more proportionate and more goal-oriented approach with less reliance on compliance and regulation as the key drivers.

The profession needs to use the tools it is being given to lead, inspire commitment, seek out ways to add value and initiate the actions that will enable organizational success. Competency measurement will bring out the differences between organizations in a more sophisticated way than regulation can achieve. The tools used by health and safety professionals, their risk assessments, their policies and their audits will be sensitive to specific organizational contexts and business challenges. A competency approach will also be more proactive in managing perceptions of legitimacy through the positive evidence of achievements. It demonstrates a profession better equipped and better positioned to contribute to tangible business outcomes. It is a new opportunity for advocacy within the profession, challenging its members to step up and respond to organizational needs and public expectations while leading debate on the priorities and directions for future health and safety investments.

Health and safety has proven itself to be a hugely dynamic discipline. It continues to evolve in response to the external factors that influence its positioning and influence. Even over the five-year period of the research programme, we have witnessed a repositioning of health and safety influenced by a changing regulatory framework, the impact of sustainability requirements and the increasing complexity of the organizational environments within which the profession operates. The old adage that good health and safety management is good business is still strong. It is, however, also one that is having an increasingly global impact. A culture of protectionism about professional knowledge and experience has begun to open up. Many organizations are rightly proud of their achievements in health and safety and more open about sharing their methods and results with others – this is a vital and positive foundation for further improvements.

The profession needs to build its future authority on a proactive approach to the collection and sharing of evidence and lessons learned in different situations (good or bad). This could be through *ad hoc* research projects or the assembly of empirical experience from within its wide network. It is of utmost importance for practitioners to stay up to date, identify leading practices and continue to share experiences with peers, to become a genuine community of practice. The emerging appreciation that failing to recognize health and safety as a key business risk will have an impact on the success of the leadership strategy and related objective of an organization, the profession has a duty to enable their organization to respond to this.

This authority of expertise is a precondition for addressing misperceptions, whether among the wider public, in the boardroom or among the workforce and supply chain. It needs to be accompanied by skills in communication to ensure that messages are shaped to the needs and capacities of audiences rather than driven by the interests and formats preferred by authors. This communication must go beyond raising awareness to promoting ownership and accountability. It also needs to be matched by listening skills, receptiveness to feedback and recognition of the contributions that others can make.

The health and safety profession is renowned for its technical expertise and capability. There are many inspiring examples of the profession mobilizing organizations to achieve their vision in the most challenging of conditions. There is increasing recognition that a culture of care will be transformational for organizations – visibly acknowledging the paramount importance of staff welfare can engender a strong commitment to the core values of the organization. The profession as a whole needs to align itself with this transformation. It is also increasingly a global profession, with access to huge volumes of data and intelligence that can be translated into practical, proportionate and risk-based proposals for organizations.

The research programme has reminded us of the dynamism of the environment in which the safety and health profession is operating. In contrasting the picture of work and organizations drawn by the contributors to this book with that encountered by the *Robens Committee*, we must also recognize that the findings presented here will, in their turn, also become obsolete. It is a frustration – and a challenge – for the social sciences, that their objects of study evolve so rapidly and must constantly be understood anew. However, the profession, and its representative bodies, face exactly the same problem. They need to grasp the vital importance of using their knowledge and experience as a source of flexibility and adaptation. Research of the kind reported here is not just a one-off exercise but part of a continuous attempt to monitor what is happening around the profession, and to respond appropriately. IOSH has, and must continue to cultivate, a rich fund of knowledge gleaned from its members and its interactions with other stakeholders such as academics, regulators, businesses, trade unions, advocacy groups and politicians. Such mutual exchanges are essential for the development of strategic thinking about the profession's future and the ways in which it will remain relevant.

While the world changes, however, one thing must remain constant. We return to the theme with which we began – to the vision of a world in which every worker completes every shift, and their whole working life, alive, in good health and without any new disability. The health and safety profession's commitment to that goal remains unwavering, even as its members apply their collective imagination, creativity and skills to the means by which it is accomplished.

# References

Aalders, M. and Wilthagen, T. 1997. Moving beyond command-and-control: reflexivity in the regulation of occupational safety and health and the environment. *Law & Policy* 19(4), pp. 415–443.

Almond, P. 2009. The dangers of hanging baskets: regulatory myths and media representations of health and safety regulation. *Journal of Law and Society* 36(3), pp. 352–375.

Almond, P. and Colover, S. 2012. Communication and social regulation: the criminalization of work-related death. *British Journal of Criminology* 52(5), pp. 997–1016.

Andreou, N.J.A. and Leka, S. 2013. The role of corporate social responsibility in improving occupational safety and health – evidence from the field. In: Jain, A. et al. (eds.). *Occupational safety & health and corporate social responsibility in Africa: repositioning corporate responsibility towards national development*. Cranfield, UK: Cranfield Press. pp. 127–142.

Antonsson, A.B. et al. 2002. *Small enterprises in Sweden: health and safety and the significance of intermediaries in preventive health and safety*. Stockholm: National Institute for Working Life.

Austin, J.E. 2000. *The collaboration challenge: how nonprofits and businesses succeed through strategic alliances*. San Francisco: Jossey-Bass.

Avolio, B.J., Walumbwa, F.O. and Weber, T.J. 2009. Leadership: current theories, research, and future directions. *Annual Review of Psychology* 60, pp. 421–449.

Baggot, R. 1989. Regulatory reform in Britain: the changing face of self-regulation. *Public Administration* 67(4), pp. 435–454.

Bain, P. 1997. Human resource malpractice: the deregulation of health and safety at work in the USA and Britain. *Industrial Relations Journal* 28(3), pp. 176–191.

Bales, R.F. and Slater, P.E. 1955. Role differentiation in small decision-making groups. In: Parsons, T. and Bales, R.F. (eds). *Family, socialization and interaction process*. Glencoe, IL: The Free Press. pp. 259–306.

Bandle, T. 2007. Tolerability of risk: the regulator's story. In: Bouder, F. et al. (eds.). *The tolerability of risk: a new framework for risk management*. London: Earthscan. pp. 93–103.

Barbeau, E. et al. 2004. Assessment of occupational safety and health programs in small businesses. *American Journal of Industrial Medicine* 45(4), pp. 371–379.

Baril-Gingras, G. et al. 2006. The contribution of qualitative analyses of occupational health and safety interventions: an example through a study of external advisory interventions. *Safety Science* 44(10), pp. 851–874.

Barling, J., Loughlin, C. and Kelloway, E.K. 2002. Development and test of a model linking safety-specific transformational leadership and occupational safety. *Journal of Applied Psychology* 87(3), 488–496.

Barrett, B. and Howells, R. 1997. *Occupational health and safety law*. London: Pitman Publishing.

Bass, B.M. 1985. *Leadership and performance beyond expectations*. New York: Free Press.

Bearfield, G.J. 2009. Achieving clarity in the requirements and practice for taking safe decisions in the railway industry in Great Britain. *Journal of Risk Research* 12(3/4), pp. 443–453.

Beck, M. and Woolfson, C. 2000. The regulation of health and safety in Britain: From old Labour to new Labour. *Industrial Relations Journal* 31(1), pp. 35–49.

Benjamin, K. and White, J. 2003. *Occupational health in the supply chain: A literature review* (HSL/2003/06). Sheffield: Health and Safety Laboratory. www.hse.gov.uk/research/hsl_pdf/2003/hsl03-06.pdf [Accessed 8 February 2016].

Berlinger, N. 2016. *Are workarounds ethical?* New York: Oxford University Press.

Better Regulation Executive (BRE). 2008. *Improving outcomes from health and safety: A report to government by the Better Regulation Executive.* Retrieved from webarchive.nationalarchives.gov.uk/20121212135622/http://www.bis.gov.uk/files/file47324.pdf [Accessed 8 February 2016].

Black, J. 2008. Constructing and contesting legitimacy and accountability in polycentric regulatory regimes. *Regulation & Governance* 2(2), pp. 137–164.

Boardman, J. and Lyon, A. 2006. *Defining best practice in corporate occupational health and safety governance.* Sudbury, UK: HSE Books.

Brace, C. et al. 2009. *Phase 2 Report: Health & safety in the construction industry: Underlying causes of construction fatal accidents.* Norwich, UK: Her Majesty's Stationery Office.

Bradshaw, L. et al. 2001. Provision and perception of occupational health in small and medium-sized enterprises in Sheffield, UK. *Occupational Medicine* 51(1), pp. 39–44.

Braithwaite J. 2002. *Restorative justice and responsive regulation.* Oxford: Oxford University Press.

Branch, A. 2005. The evolution of the European Social Dialogue towards greater autonomy: challenges and potential benefits. *International Journal of Comparative Law and Industrial Relations* 21(2), pp. 321–346.

Brazier, A., Gait, A. and Waite, P. 2004. *Different types of supervision and the impact on safety in the chemical and allied industries* (part 2 of 3). Sudbury, UK: HSE Books.

BSI. 1996. *Guide to occupational health and safety management systems.* BS8800:1996. London: BSI.

BSI. 1999. *Occupational health and safety management systems – specification.* OHSAS18001. London: BSI.

BSI. 2004. *Guide to occupational health and safety management systems.* BS8800:2004. London: BSI.

BSI. 2008a. *Guide to achieving effective occupational health and safety performance.* BS18004:2008. London: BSI.

BSI. 2008b. *Occupational health and safety management systems – guidelines for the implementation of OHSAS 18001:2007.* London: BSI.

Buchanan, D.A. and Denyer, D. 2015. What's the problem? In: Denyer D. and Pilbeam, C. (eds.) *Managing change in extreme contexts.* London: Routledge. pp. 3–23.

Buckle, P. et al. 2006. Patient safety, systems design and ergonomics. *Applied Ergonomics* 37(4), pp. 491–500.

Budworth, N. and Khan, S. 2000. The continuous improvement model. Securing health together (programme 2). http://web.archive.org/web/20090714000517/http://www.hse.gov.uk/sh2/pags/continuousimprovement.htm [Accessed 8 February 2016].

Busby, J.S. and Collins, A. 2009. *Risk leadership and organisational type.* HSE. www.hse.gov.uk/research/rrpdf/rr756.pdf [Accessed 27 September 2016].

Business Dictionary. 2015. *Networked organisation definition,* www.businessdictionary.com/definition/network-organization [Accessed 8 February 2016].

Callaghan, B. 2007. Employment relations: the heart of health and safety. *Warwick Papers in Industrial Relations*. Coventry: Warwick Business School. www2.warwick.ac.uk/fac/soc/wbs/research/irru/wpir/wpir84.pdf [Accessed 8 February 2016].

Canadian Institutes for Health Research (CIHR). 2006. *Moving population and public health knowledge into action. A casebook of knowledge translation stories*. Ottawa: Canadian Institute for Health Information.

Carillo, R.A. 2005. Safety leadership: managing the paradox. *Professional Safety* July 2005: pp. 31–34.

Carlile, P.R. and Rebentisch, E.S. 2003. Into the black box: The knowledge transformation cycle. *Management Science* 49(9), pp.1180–1195.

Carson, W.G. 1982. *The other price of Britain's oil*. Oxford: Martin Robertson.

CEC. 1989. Council Directive of 12 June 1989 on the introduction of measures to encourage improvements in the safety and health of workers at work (89/291/EEC). http://eur-lex.europa.eu/legal-content/EN/TXT/PDF/?uri=CELEX:31989L0391&from=en [Downloaded 26 June 2015].

Champoux, D. and Brun, J. 2003. Occupational health and safety management in small size enterprises: an overview of the situation and avenues for intervention and research. *Safety Science* 41(4), pp. 301–318.

Chen, S. and Bouvain, P. 2009. Is corporate responsibility converging? A comparison of corporate responsibility reporting in the USA, UK, Australia, and Germany. *Journal of Business Ethics* 87(1), pp. 299–317.

Cheyne, A. et al. 2012. *Talk the talk – walk the walk: An evaluation of Olympic Park safety initiatives and communication IOSH Research Report 12.1*. Leicester, UK: Institute of Occupational Safety and Health. www.iosh.co.uk/~/media/Documents/Books%20and%20resources/Published%20research/Talk_the_talk_Walk_the_walk.pdf?la=en [Accessed 8 February 2016].

Clark, D. 2004. *The continuum of understanding*. www.nwlink.com/~donclark/performance/understanding.html [Accessed 14 March 2016].

Clarke, S. and Ward, K. 2006. The role of leader influence tactics and safety climate in engaging employees' safety participation. *Risk Analysis* 26(5), pp. 1175–1185.

Collins, H. M. 1993. The structure of knowledge. *Social Research* 60(1), pp. 95–116.

COMAH. 2011. *Buncefield: why did it happen?* Sudbury, UK: HSE.

Conchie, S. and Donald, I. 2009. The moderating role of safety-specific trust on the relation between safety-specific leadership and safety citizenship behaviors. *Journal of Occupational Health Psychology* 14(2), pp. 137–147.

Conchie, S. and Moon, S. 2010. *Promoting active safety leadership*. Leicester, UK: IOSH.

Conchie, S., Taylor, P. and Donald, I. 2012. Promoting safety voice with safety-specific transformational leadership: The mediating role of two dimensions of trust. *Journal of Occupational Health Psychology* 17(1), pp. 105–115.

Conchie, S. Moon, S. and Duncan, M. 2013. Supervisors' engagement in safety leadership: factors that help and hinder. *Safety Science* 51(1), pp. 109–117.

Conzola, V.C. and Wogalter, M.S. 2001. A communication–human information processing (C–HIP) approach to warning effectiveness in the workplace. *Journal of Risk Research* 4(4), pp. 309–322.

Cooper, M.D. 2000. Toward a model of safety culture. *Safety Science* 35(2), pp. 111–136.

Corr Willbourn. 2009. *Report of qualitative research amongst 'hard to reach' small construction site operators* (HSE Research Report 719). Norwich, UK: Her Majesty's Stationery Office.

Crawford, J. O. et al. 2016. Evaluation of knowledge transfer for occupational safety and health in an organisational context: development of an evaluation framework. *Policy and practice in health and safety*. In press.

Crombie, K. F. 2000. Deregulation of health and safety laws in the USA and UK: Past practices, recent trends and future options. *Corporate Crime and Governance*. Retrieved from: www.scottishlaw.org.uk/journal/oct2000/corpcrimdis.pdf [Accessed 27 September 2016].

Cross, R. et al. 2007. *The role of networks in organizational change*. Washington, DC: McKinsey & Company.

Crown. 1967. *Annual Report of HM Chief Inspector of Factories* (Cmnd 3745). London: HMSO.

Crown. 1988. *The work of the Health and Safety Commission and Executive* (1987/88 HC 267), minutes of evidence 20 January 1988. London: HMSO.

Crown. *c*.1969. Memorandum from Chief Inspector of Factories to managing directors and managers of firms employing more than 50 people. Found at MRC MSS.292D/146/18/1.

Cullen, Lord. 1990. *The Public Inquiry into the Piper Alpha Disaster*. London: HMSO.

Cullen, P.C. 2001. *The Ladbroke Grove rail inquiry*. Part 2 Report. Sudbury, UK: HSE Books.

Culyer, A.J. et al. 2008. What is a little more health and safety worth? In: Tompa, E., Culyer, C.J. and Dolinschi, R. (eds.). *Economic evaluation of interventions for occupational health and safety*. Oxford: Oxford University Press, pp.15–35.

Cummings, R. 2006. Expert views on the evidence base for effective health and safety management (phase 2). HSL/2006/109. HSL. www.hse.gov.uk/research/hsl_pdf/2006/hsl06109.pdf [Accessed 27 September 2016].

Currie, G. and Lockett, A. 2011. Distributing leadership in health and social care: concertive, conjoint or collective? *International Journal of Management Reviews* 13(3), pp. 286–300.

Dahl, O. and Olsen, E. 2013. Safety compliance on offshore platforms: a multi-sample survey on the role of perceived leadership involvement and work climate. *Safety Science* 54(1), pp. 17–26.

Dalton, A.J.P. 1991. *Health and safety: An agenda for change*. London: WEA.

Dalton, A.J.P. 1992. Lessons from the United Kingdom: fightback on workplace hazards, 1979–1992. *International Journal of Health Services* 22(3), pp. 489–495.

Dalton, A.J.P. 1998. *Safety, health and environmental hazards at the workplace*. London: Cassell.

Danish Commerce and Companies Agency. 2011. *Corporate social responsibility and reporting in Denmark: Impact of the second year subject to the legal requirements for reporting on CSR in the Danish Financial Statements Act*. Retrieved from http://csrgov.dk/file/319199/corporate_social_responsibility_and_reporting_in_denmark_november_2011.pdf.pdf [Accessed 27 September 2016].

Davenport, T.H. and Prusak, L. 1998. *Working knowledge: How organizations manage what they know*. Cambridge, MA: Harvard Business Press.

Davidson, N. 2006. *Davidson Review – Final Report. The Stationery Office*. Retrieved from: http://webarchive.nationalarchives.gov.uk/20090609003228/http://www.berr.gov.uk/files/file44583.pdf [Accessed 27 September 2016].

Dawson, D. et al. 1988. *Safety at work: The limits of self-regulation*. Cambridge: Cambridge University Press.

Dekker, S. 2003. Failure to adapt or adaptations that fail: contrasting models on procedures and safety. *Applied Ergonomics* 34(3), pp. 233–238.

Deming, W.E. 1982. *Quality, productivity, and competitive position*. Cambridge, MA: Massachusetts Institute of Technology.

Denis, J–L., Langley, A. and Sergi, V. 2012. Leadership in the plural. *The Academy of Management Annals* 6(1), pp. 211–283.

Denison, D.R. and Spreitzer, G.M. 1991. Organizational culture and organizational development: A competing values approach. *Research in Organizational Change and Development* 5(1), pp. 1–21.

Denning, J. 1985. The hazards of women's work. *New Scientist* 17 January, pp. 12–15.

Department of Employment. 1986. *Building businesses not barriers*. Government White Paper, Cm. 9794, London: HMSO.

Department of Environment. 1985. *Lifting the burden*. Government White Paper, Cm.9571. London: HMSO.

Department of Transport (DfT). 1987. The Merchant Shipping Act 1894 mv *Herald of Free Enterprise*. Report of Court No. 8074. Formal Investigation. London: HMSO.

Department of Transport (DfT). 1988. Investigation into the King's Cross Underground Fire. London: HMSO.

Drennan, F.S. and Richey, D. 2012. Skills-based leadership: the first line supervisor – part II. *Professional Safety* March 2012, pp. 50–54.

Dunlop, C. 2014 *Health and safety myth-busters challenge panel: case analysis*. Research Paper, University of Exeter.

Eakin, J. 1992. Leaving it up to the workers: sociological perspective on the management of health and safety in small workplaces. *International Journal of Health Services* 22(4), pp. 689–704.

Eakin, J. and MacEachen, E. 1998. Health and the social relations of work: a study of the health-related experiences of employees in small workplaces. *Sociology of Health & Illness* 20(6), pp. 896–914.

Eurofound, 2009. *Social contract*. Retrieved from: www.eurofound.europa.eu/emire/UNITED%20KINGDOM/SOCIALCONTRACT-EN.htm

European Agency for Safety and Health at Work (EU-OSHA). 2009. *OSH in figures: stress at work – facts and figures*. Luxembourg: Office for Official Publications of the European Communities.

European Agency for Safety and Health at Work (EU-OSHA). 2012. *Drivers and barriers for psychosocial risk management: An analysis of the findings of the European Survey of Enterprises on New and Emerging Risks* (ESENER). Luxembourg: Office for Official Publications of the European Communities.

European Agency for Safety and Health at Work (EU-OSHA). 2013. *Occupational safety and health and education: A whole school approach*. Luxembourg: Publications Office of the European Union.

European Agency for Safety and Health at Work (EU-OSHA). [no date]. *Rome declaration on mainstreaming OSH into education and training*. Retrieved from: https://osha.europa.eu/en/topics/osheducation/rome.stm [Accessed 27 September 2016].

European Commission (EC). 2003. *The new SME definition. User guide and model declaration*. European Commission. Luxembourg: Office for Official Publications of the European Communities.

European Commission (EC). 2007. *Communication on the Community strategy 2007–2012 on health and safety at work* COM(2007)62. Retrieved from: http://eur-lex.europa.eu/LexUriServ/LexUriServ.do?uri=CELEX:52007DC0062:EN:NOT [Accessed 27 September 2016].

Evans, D.D., Michael, J.H., Wiedenbeck, J.K. and Ray, C.D. 2005. Relationships between organizational climates and safety-related events at four wood manufacturers. *Forest Products Journal* 55(6), pp. 23–28.

Eves, D. 2014. History of occupational safety and health: 'Two steps forward, one step back'. www.historyofosh.org.uk/brief/index.html [Accessed 26 June 2015].

Facilities (Anonymous). 1993. Health and safety: 1993. *Facilities* 11(7), pp. 11–17.

Fairman, R. 1994. Robens – 20 years on. *Health and Safety Information Bulletin* 221, pp.13–16.

Fairman, R. and Yapp, C. 2004. Compliance with food safety legislation in small and micro-businesses: enforcement as an external motivator. *Journal of Environmental Health Research* 3(2), pp. 44–52.

Finneran, A. et al. 2012. Learning to adapt health and safety initiatives from mega projects: an Olympic case study. *Policy and Practice in Health and Safety* 10(2), pp. 81–102.

Fitzsimmons, D., Turnbull James, K. and Denyer, D. 2011. Alternative approaches for studying shared and distributed leadership. *International Journal of Management Reviews* 13(3), 313–328.

Fleming, M. 1999. *Effective supervisory safety leadership behaviours in the offshore oil and gas industry.* HSE. www.hse.gov.uk/research/otopdf/1999/oto99065.pdf [Accessed 27 September 2016].

Fonteyn, P. et al. 1997. Small business owners' knowledge of their occupational health & safety (OHS) legislative responsibilities. *International Journal of Occupational Safety and Ergonomics* 3(1–2), pp. 41–58.

Forck, M. 2012. Why safety leadership is hard: 7 secrets to help you succeed. *Professional Safety* October, pp. 34–35.

Fuller, C. and Vassie L. 2005. *Benchmarking employee supervisory processes in the chemical industry.* HSE. www.hse.gov.uk/research/rrpdf/rr312.pdf [Accessed 27 September 2016].

Gadd, S. and Collins, A.M. 2002. *Safety culture: A review of the literature.* HSL/2002/25. HSE. www.hse.gov.uk/research/hsl_pdf/2002/hsl02-25.pdf [Accessed 27 September 2016].

Geller, E S. 2000. 10 leadership qualities for a total safety culture. *Professional Safety* May, pp. 38–41.

Geller, E S. 2008. People based leadership: enriching a work culture for world-class safety. *Professional Safety* March, pp. 29–36.

Genn, H. 1993. Business responses to the regulation of health and safety in England. *Law & Policy* 15(3), pp. 219–233.

Gherardi, S. and Nicolini, D. 2002. Learning the trade: A culture of safety in practice. *Organization* 9(2), pp. 191–223.

Gibb, A.G.F. et al. 2016a. *Engagement of micro, small and medium-sized enterprises in occupational safety and health.* Leicester, UK: Institution of Occupational Safety and Health.

Gibb, A.G.F. et al. 2016b. *Occupational safety and health in networked organisations.* Leicester, UK: Institution of Occupational Safety and Health.

Goldman, L. and Lewis, J. 2004. How safe is safe enough? *Occupational Health* 56(6), pp. 12–14.

Goodwin, C., 1994. Professional vision. *American Anthropologist* 96(3), pp. 606–633.

Gordon R. D. 2002. Conceptualizing leadership with respect to its historical-contextual antecedents to power. *The Leadership Quarterly* 13(2), pp. 151–167.

Graham, I.D. et al. 2006. Lost in knowledge translation: time for a map? *Journal of Continuing Education in the Health Professions* 26(1), pp. 13–24.

Grint, K. 2000. *The arts of leadership.* Oxford: Oxford University Press.

Gronn P. 2002. Distributed leadership as a unit of analysis. *The Leadership Quarterly* 13(4), pp. 423–451.

Guardian. 1974. The holocaust at Flixborough. *Guardian* 3 June, p. 10.

Guidotti, T.L. et al. 2000. The Fort McMurray demonstration project in social marketing: theory, design, and evaluation. *American Journal of Preventive Medicine* 18(2), pp. 163–169.

Gulick, L. H. 1936. Notes on the theory of organization. In: Gulick, L. and Urwick, L. (eds.). *Papers on the Science of Administration.* New York: Institute of Public Administration, pp. 3–35.

Gunningham, N. et al. 2004. Social license and environmental protection: why businesses go beyond compliance. *Law & Social Inquiry* 29(2), pp. 307–341.

Hackitt, J. 2015. Gold plating or feather bedding? Blog post, 31 July. Retrieved from: www.hse.gov.uk/news/judith-risk-assessment/gold-plating-or-feather-bedding310715. htm [Accessed 27 February 2016].

Hale, A. and Borys, D. 2013. Working to rule, or working safely? Part 1: State of the art review – Part 2: The management of safety rules and procedures. *Safety Science* 55(June), pp. 207–231.

Hales, C.P. 1999. Why do managers do what they do? Reconciling evidence and theory in accounts of managerial work. *British Journal of Management* 10(4), pp. 335–350.

Harvey, D. 2005. *A brief history of neoliberalism.* Oxford: Oxford University Press.

Hasle, P. and Limborg, H.J. 2006. A review of the literature on preventive occupational health and safety activities in small enterprises. *Industrial Health* 44(1), pp. 6–12.

Hasle, P. et al. 2009. Small enterprise owners' accident causation attribution and prevention. *Safety Science* 47(1), pp. 9–19.

Hasle, P. et al. 2012. The working environment in small firms: Responses from owner-managers. *International Small Business Journal* 30(6), pp. 622–639.

Healey, N. and Sugden, C. 2012. *Safety culture on the Olympic Park.* HSE. www.hse.gov.uk/research/rrhtm/rr942.htm [Accessed 27 September 2016].

Health and Safety at Work etc. Act 1974. www.legislation.gov.uk/ukpga/1974/37/pdfs/ukpga_19740037_en.pdf [Downloaded 26 June 2015].

Heifetz, R.A. and Laurie, D.L. 1997. The work of leadership. *Harvard Business Review* 75(1), pp. 124–134.

Henshaw, J.L. et al. 2007. The employer's responsibility to maintain a safe and healthful work environment: A historical review of societal expectations and industrial practices. *Employee Responsibilities and Rights Journal* 19(3), pp. 173–173.

Hollnagel, E. 2014. *Safety-I and Safety-II: the past and future of safety management.* Farnham, UK: Ashgate.

Holmes, N. and Gifford, S. 1997 Narratives of risk in occupational health and safety: why the 'good' boss blames his tradesman and the 'good' tradesman blames his tools. *Australian and New Zealand Journal of Public Health* 21(1), pp. 11–16.

Holmes, N. et al. 2000. An exploratory study of meanings of risk control for long term and acute effect occupational health and safety risks in small business construction firms. *Journal of Safety Research* 30(4), pp. 251–261.

Hood, C. 1991. A public management for all seasons? *Public Administration* 69(1), pp. 3–19.

HSC. 1994. *Review of health and safety regulation – Main report.* London: HSE Books.

HSC. 2000. *The Southall Rail Accident Inquiry Report.* Norwich, UK: HMSO.

HSC. 2004a. *A strategy for workplace health and safety in Great Britain to 2010 and beyond.* Health and Safety Executive. HSE Books: Sudbury, Suffolk.

HSC. 2004b. *Proposals for a public consultation campaign to support the development of Management Standards to tackle work-related stress.* www.hse.gov.uk/aboutus/meetings/hscarchive/2004/060404/c05.pdf [Accessed 27 September 2016].

HSE. 1985. *Report by HM Chief Inspector of Factories 1985.* London: HMSO.

HSE. 1991a. *It's your job to manage safety.* INDG103(L). Health and Safety Executive. HSE Books: Sudbury, Suffolk.

HSE. 1991b. *Successful health and safety management.* (HSG65). First Edition. Health and Safety Executive. HSE Books: Sudbury, Suffolk.

HSE. 1992a. *The tolerability of risks from nuclear power stations.* HSE Books: Sudbury, Suffolk.

HSE. 1992b. *The fire at Hickson & Welch Ltd. A report of the investigation by the Health and Safety Executive into the fatal fire at Hickson & Welch Ltd, Castleford on 21 September 1992.* Sudbury, UK: HSE Books.

HSE. 1993. *ACSNI Study Group on human factors. Third Report: Organising for safety.* HSE Books: Sudbury, Suffolk.

HSE. 1994a. *The chemical release and fire at the Associated Octel Company Limited.* Sudbury, UK: HSE Books.

HSE. 1994b. *The explosion and fires at the Texaco Refinery, Milford Haven 24 July 1994. A report of the investigation by the Health and Safety Executive into the explosion and fires on the Pembroke Cracking Company Plant at the Texaco Refinery, Milford Haven on 24 July 1994.* Sudbury, UK: HSE Books.

HSE. 1997. *Successful health and safety management* (HSG65). Second Edition. Health and Safety Executive. HSE Books: Sudbury, Suffolk.

HSE. 2002. *Directors' responsibilities for health and safety* INDG343. HSE Books: Sudbury, Suffolk.

HSE. 2004. *Leadership for the major hazard industries.* INDG277(rev). Health and Safety Executive. http://www.hse.gov.uk/pubns/indg277.pdf [Accessed 27 September 2016].

HSE. 2006. *HSC/E Simplification plan 2006.* Retrieved from: www.hse.gov.uk/simplification/simplification06.pdf [Accessed 27 September 2016].

HSE. 2008. *Employers' Liability (Compulsory Insurance) Act 1969: A Guide for Employers.* Retrieved from www.hse.gov.uk/pubns/hse40.pdf [Accessed 27 September 2016].

HSE. 2009. *The health and safety of Great Britain – Be part of the solution.* Retrieved from: http://www.hse.gov.uk/aboutus/strategiesandplans/strategy09.pdf [Accessed 27 September 2016].

HSE. 2012. *Leadership and worker involvement toolkit. Good health and safety leadership.* Health and Safety Executive. http://www.hse.gov.uk/construction/lwit/assets/downloads/good-health-safety-leadership.pdf [Accessed 27 September 2016].

HSE. 2013. *Managing for health and safety* (HSG65). Third Edition. Health and Safety Executive. http://www.hse.gov.uk/pubns/books/hsg65.htm [Accessed 27 September 2016].

Hutchins, E. 1995. *Cognition in the Wild.* Cambridge, MA: MIT Press.

Hutter, B.M. 2005. *The attractions of risk-based regulation: Accounting for the emergence of risk ideas in regulation.* London: London School of Economics and Political Science.

ICAEW. 1999. *Internal control: Guidance for Directors on the Combined Code.* London: Institute of Chartered Accountants in England and Wales.

ILO/WHO. 1950. *Occupational health.* Geneva: Joint ILO/WHO Committee on Occupational Health.

Institution of Occupational Safety and Health (IOSH). 2010. *Getting the balance right: Institution of Occupational Safety and Health response to the Young Report, 'Common sense, common safety'.* Leicester: Institution of Occupational Safety and Health.

Institution of Occupational Safety and Health (IOSH). 2012. *More haste, less speed? In brief: IOSH's response to the Löfstedt Review.* Leicester: Institution of Occupational Safety and Health.

Jain, A. and Leka, S. 2015. *Occupational health and safety legitimacy in the UK: A review of quantitative data.* Leicester: Institution of Occupational Safety and Health.

Jain, A. et al. 2011. Corporate social responsibility and psychosocial risk management in Europe. *Journal of Business Ethics* 101(4), pp. 619–633.

James, P. and Walters, D. 1997. Non-union rights of involvement: The case of health and safety at work. *Industrial Law Journal* 26(1), pp. 35–50.

James, P. and Walters, D. 2002. Worker representation in health and safety: options for regulatory reform. *Industrial Relations Journal* 33(2), pp. 141–156.

James, P. et al. 2004. The use of external sources of health and safety information and advice: the case of small firms. *Policy and Practice in Health and Safety* 2(1), pp. 91–104.

Johnstone, R. and Carson, W.G. 2002. Occupational health and safety, Regulation of. *International Encyclopedia of the Social & Behavioral Sciences.* pp. 10835–10839.

Kang, J. et al. 2010. Revisiting knowledge transfer: Effects of knowledge characteristics on organizational effort for knowledge transfer. *Expert Systems with Applications* 37(12), pp. 8155–8160.

Kankaanpää, E. et al. 2008. Economics for occupational safety and health. *Scandinavian Journal of Work Environment & Health Supplements* 34(5), pp. 9–13.

Kapp E. 2012a. The influence of supervisor leadership practices and perceived group safety climate on employee safety performance. *Safety Science* 50(4), pp. 1119–1124.

Kapp, E.A. 2012b. Leadership alone is not enough. *Professional Safety* May, p. 10.

Kelloway, E.K., Mullen, J. and Francis L. 2006. Divergent effects of transformational and passive leadership on employee safety. *Journal of Occupational Health Psychology* 11(1), pp. 76–86.

Kemp, R.V. 1991. Risk tolerance and safety management. *Reliability Engineering & System Safety* 31(3), pp. 345–353.

King, K., Lunn, S. and Michaelis, C. 2010. *Director leadership behaviour research.* HSE. www.hse.gov.uk/research/rrpdf/rr816.pdf [Accessed 27 September 2016].

Krause, T.R. and Weekley, T. 2005. Safety leadership: a four factor model for establishing a high-functioning organization. *Professional Safety* November, pp. 34–40.

Ladkin, D. 2010. *Rethinking leadership: A new look at old leadership questions.* Cheltenham, UK: Edward Elgar.

Laroche. E. and Amara, N. 2011. Transfer activities among Canadian researchers: Evidence in occupational safety and health. *Safety Science* 49(3), pp. 406–415.

Lavack, A.M. et al. 2008. Enhancing occupational health and safety in young workers: The role of social marketing. *International Journal of Nonprofit and Voluntary Sector Marketing* 13(3), pp, 193–204.

Lave, J. and Wenger, E. 1991. *Situated learning: Legitimate peripheral participation.* Cambridge: Cambridge University Press.

Layfield, F. 1987. *Sizewell B Public Inquiry: Report by Sir Frank Layfield.* London: HMSO.

Leathley, B. 2013. *What is competence?* Retrieved from: www.healthandsafetyatwork.com/hsw/risk-assessment/competence [Accessed 27 September 2016].

Leka, S. et al. 2006. *Exploring health and safety practitioners training needs in workplace health issues.* Leicester, UK: Institution of Occupational Safety and Health.

Leka, S. et al. 2010. Policy-level interventions and work-related psychosocial risk management in the European Union. *Work & Stress* 24(3), pp. 298–307.

Leka, S. et al. 2011. The role of policy for the management of psychosocial risks at the workplace in the European Union. *Safety Science* 49(4), pp. 558–564.

Lekka, C. and Healey, N. 2012. A review of the literature on effective leadership behaviours for safety. HSE. www.hse.gov.uk/research/rrhtm/rr952.htm [Accessed 27 September 2016].

Leonard, P.E. and Leonard, L.J. 2001. The collaborative prescription: Remedy or reverie? *International Journal of Leadership in Education* 4(4), pp. 383–399.

Lin, M. and Li, N. 2010. Scale-free network provides an optimal pattern for knowledge transfer. *Physica A: Statistical Mechanics and its Applications* 389(3), pp. 473–480.

Löfstedt, R.E. 2004. The swing of the regulatory pendulum in Europe: From precautionary principle to (regulatory) impact analysis. *The Journal of Risk and Uncertainty* 28(3), pp. 237–260.

Löfstedt, R.E. 2011a. *Reclaiming health and safety for all: An independent review of health and safety legislation.* Department for Work and Pensions. London: HM Stationery Office.

Löfstedt, R.E. 2011b. Risk versus hazard – how to regulate in the 21st century. *The European Journal of Risk Regulation* 2011(2), pp. 149–168.

Luria, G., Zohar, D. and Erev, I. 2008. The effect of workers' visibility on effectiveness of intervention programs: Supervisory-based safety interventions. *Journal of Safety Research* 39(3), pp. 273–280.

Lyddon, D. 2010. *From Gowers to Robens: health and safety reform in the UK, 1945–74.* European Social Science History Conference, Ghent, 13–16 April, 2010. https://esshc.socialhistory.org/esshc-user/programme/2010?day=20&time=94&session=2131 [Accessed 8 February 2016].

Lyddon, D. 2012. *Britain at work: Voices from the workplace 1945–1995.* Retrieved from: www.unionhistory.info/britainatwork/narrativedisplay.php?type=healthandsafety [Accessed 27 September 2016].

Mackay, C.J. et al. 2004. Management standards and work-related stress in the UK: policy background and science. *Work & Stress* 18(2), pp. 91–112.

The Management of Health and Safety at Work Regulation. 1992. www.legislation.gov.uk/uksi/1992/2051/contents/made [Downloaded 26 June 2015].

Mascini. P. 2005. The blameworthiness of health and safety rule violations. *Law & Policy* 27(3), pp. 472–490.

Mathis T. 2013. Be an effective safety leader. *Professional Safety* January, p. 19.

Mayrowetz, D. 2008. Making sense of distributed leadership: exploring the multiple usages of the concept in the field. *Educational Administration Quarterly* 44(3), pp. 424–435.

Meyer, M. 2010. The rise of the knowledge broker. *Science Communication* 32(1), pp.118–127.

Moore, M. 1995. *Creating public value: Strategic management in government.* Cambridge, MA: Harvard University Press.

Moran, M. 2003. *The British regulatory state: High modernism and hyper innovation.* Oxford: Oxford University Press.

Mullen, J.E. and Kelloway, E.K. 2009. Safety leadership: a longitudinal study of the effects of transformational leadership on safety outcomes. *Journal of Occupational and Organizational Psychology* 82(2), pp. 253–272.

Murray, S. R. and Peyrefitte, J. 2007. Knowledge type and communication media choice in the knowledge transfer process. *Journal of Managerial Issues* 19(1), pp. 111–133.

Neal, A.C. 2004. *Providing "teeth" for the right to a safe and healthy working environment?* The Hague: Kluwer Law International.

Nonaka, I. 1994. A dynamic theory of organizational knowledge creation. *Organization Science,* 5(1), pp.14–37.

Nonaka, I. and Takeuchi, H. 1995. *The knowledge-creating company: How Japanese companies create the dynamics of innovation.* Oxford: Oxford University Press.

Nonaka, L., Takeuchi, H. and Umemoto, K. 1996. A theory of organizational knowledge creation. *International Journal of Technology Management* 11(7–8), pp. 833–845.

O'Connell, R. 2004. Making the case for OHSAS 18001. *Occupational Hazards* 66(6), pp. 32–33.

O'Dea, A. and Flin, R. 2003. *The role of managerial leadership in determining workplace safety outcomes.* HSE. www.hse.gov.uk/research/rrhtm/rr044.htm [Accessed 27 September 2016].

Parker, D. et al. 2007. A comparison of the perceptions and beliefs of workers and owners with regard to workplace safety in small metal fabrication businesses. *American Journal of Industrial Medicine* 50(12), pp. 999–1009.

Parker, D. et al. 2012. A qualitative evaluation of owner and worker health and safety beliefs in small auto collision repair shops. *American Journal of Industrial Medicine* 55(5), pp. 474–482.

Perrow, C. 1981. Normal accident at Three Mile Island. *Society* 18(5), pp.17–26.

Petersen D. 2004. Leadership and safety excellence: a positive culture drives performance. *Professional Safety* October, pp. 28–32.

Pilbeam, C., Davidson, D., Doherty, N. and Denyer, D. 2015. *Safe working: Designing safety practices for service sector organizations*. Research Report for IOSH. Leicester: IOSH.

Pink, S. and Morgan, J. 2013. Short term ethnography: intense routes to knowing. *Symbolic Interaction* 36(3), pp. 351–361.

Pink, S. et al. 2014a. Safety in movement: mobile workers, mobile media. *Mobile Media and Communication* 2(3), pp. 335–351.

Pink, S. et al. 2014b. Managing the safe hand: gels, water, gloves and the materiality of tactile knowing. *The Journal of Material Culture* 19(4), pp. 425–442.

Plomp, H.N. 2008. The impact of the introduction of market incentives on occupational health services and occupational health professionals: Experiences from The Netherlands. *Health Policy* 88(1), pp. 25–37.

Policy and Strategic Projects Division. 2002. *Victorian nurses back injury prevention project: Evaluation report*. Melbourne: Victorian Government Department of Human Services.

Poxon, B., Coupar, W., Findlay, J., Luckhurst, D., Stevens, R. and Webster, J. 2007. *Using soft people skills to improve worker involvement in health and safety*. HSE. www.hse.gov.uk/ research/rrpdf/rr580.pdf [Accessed 27 September 2016].

Rabinowitz, R.S. 2002. *Occupational safety and health law*. Washington, DC: Bureau of National Affairs.

Rimington, J. et al. 2003. *Application of risk based strategies to workers' health and safety protection: UK experience*. The Ministry of Social Affairs and Employment (SZW). Retrieved from: www.risk-support.co.uk/SZW-published_report.pdf [Accessed 27 September 2016].

Robens, Lord. 1972. *Safety and Health at Work: Report of the Committee 1970–72 [The Robens Report]*. London: HMSO.

Robinson, A.M. and Smallman, C. 2006. The contemporary British workplace: A safer and healthier place? *Work, Employment & Society* 20(1), pp. 87–107.

Rogers, E.M. 1983. *Diffusion of innovations*. Third ed. New York: Free Press.

Roy, M. et al. 2003. Knowledge networking: a strategy to improve workplace health and safety knowledge transfer. *Electronic Journal on Knowledge Management* 1(2), pp. 159–166.

Sacks, H. 1972. Notes on police assessment of moral character. In: Sudnow, D. (ed.). *Studies in social interaction*. New York: Free Press. pp. 280–293.

Safety and Health Practitioner. 2010. *HSE cuts could work against Government's welfare ambitions*. Retrieved from www.shponline.co.uk/hse-cuts-could-work-against-government-s-welfare-ambitions [Accessed 27 September 2016].

Sarat, A. and Felstiner, W.L.F. 1989. Lawyers and legal consciousness: law talk in the divorce lawyer's office. *Yale Law Journal* 98(8), pp. 1663–1688.

Senapathi, R. 2011. Dissemination and utilisation: Knowledge. *SCMS Journal of Indian Management* 8(2), pp. 85–105.

Simpson, B. 1977. Transcript of opening remarks, first open meeting of the Advisory Committee on Asbestos, 27 June.

Smallman, C. and John, G. 2001. British directors' perspectives on the impact of health and safety on corporate performance. *Safety Science* 38(3), pp. 227–239.

Sørensen, O. H. et al. 2007. Working in small enterprises – is there a special risk? *Safety Science* 45(10), pp. 1044–1059.

Spangenberg, S. et al. 2002. The construction of the Oresund link between Denmark and Sweden: The effect of a multi-faceted safety campaign. *Safety Science* 40(5), pp. 457–465.

Spare, P. 2003. Need for improved safety priorities. Letter in *The Times*, 22 August, p. 23.

Spiers, C. 2003. Tools to tackle workplace stress. *Occupational Health* 55(12), pp. 22–25.

Spillane, J.P. 2006. *Distributed leadership*. San Francisco: Jossey-Bass.

Suchman, M. 1995. Managing legitimacy: strategic and institutional approaches, *The Academy of Management Review* 20(3), pp. 571–610.

Suddaby, R. 2010. Construct clarity in theories of management and organization. *Academy of Management Review* 35(3), pp. 346–357.

Tang, F. et al. 2006. Estimating the effect of organizational structure on knowledge transfer: A neural network approach. *Expert Systems with Applications* 30(4), pp. 796–800.

Tessier, S., 2013. Un contrôle de gestion basé sur la culture: le cas de la société Timpson. *Gestion* 38(1), pp. 56–65.

Thorpe R., Gold J. and Lawler, J. 2011. Locating distributed leadership. *International Journal of Management Reviews* 13(3), pp. 239–250.

Timpson, J. 2010. *Upside down management: A common sense guide to better business*. Chichester, UK: John Wiley & Sons.

Tombs, S. 1996. Injury, death, and the deregulation fetish: The politics of occupational safety regulation in UK manufacturing industries. *International Journal of Health Services* 26(2), pp. 309–329.

Tubby, J. 1979. Safety Act not good enough. Letter in *The Land Worker*. April, p. 6.

van Dijk, F.J.H. et al. 2010. A knowledge infrastructure for occupational safety and health. *Journal of Occupational and Environmental Medicine* 52(12), pp. 1262–1268.

Vecchio-Sadus, A.M. and Griffiths, S. 2004. Marketing strategies for enhancing safety culture. *Safety Science* 42(7), pp. 601–619.

Vickers, I. et al. 2005. Understanding small firm responses to regulation: the case of workplace health & safety. *Policy Studies* 26(2), pp. 149–169.

Walker, M. 2005. Unpacking the nature of demand and supply relationships in the mining capital goods and services cluster: the case of PGMs. In *Annual Forum 2005: trade and uneven development: opportunities and challenges*. Johannesburg, Pretoria: Trade & Industrial Policy Strategies. www.tips.org.za/files/769.pdf [Accessed 8 February 2016].

Walls, J. et al. 2004. Critical trust: understanding lay perceptions of health and safety risk regulation. *Health, Risk, and Society* 6(2), pp. 133–150.

Walters, D. 2006. One step forward, two steps back: worker representation and health and safety in the United Kingdom. *International Journal of Health Services* 36(1). pp. 87–111.

Walters, D. and James, P. 2009. *Understanding the role of supply chains in influencing health and safety at work*. Leicester, UK: Institution of Occupational Safety and Health.

Weick, K.E. 1987. Organizational culture as a source of high reliability. *California Management Review* 24(2), pp. 112–127.

Weick, K.E. and Roberts, K.H. 1993. Collective mind in organizations: heedful interrelating on flight decks. *Administrative Science Quarterly* 38(3), pp. 357–381.

Williams J.H. 2002. Improving safety leadership: using industrial/organizational psychology to enhance safety performance. *Professional Safety* April, pp. 43–47.

Williams, P. 2002. The competent boundary spanner. *Public Administration*, 80(1), pp. 103–124.

Winkler, C. 2006. *Client/contractor relationships in managing health and safety on projects.* RR462. Retrieved from: www.hse.gov.uk/research/rrpdf/rr462.pdf [Accessed 27 September 2016].

Wolman, H. 1981. The determinants of program success and failure. *Journal of Public Policy* 1(4), pp. 433–464.

Wright, M. 1998. *Factors motivating proactive health and safety management.* Contract Research Report 179. Sudbury, UK: HSE Books.

Wright, M., and Marsden, S. 2005. *A response to the CCA report 'making companies safe: what works?' Research Report 332.* Sudbury, UK: HSE Books.

Wright, M. et al. 2004. *Building an evidence base for Health and Safety Commission Strategy to 2010 and beyond: A literature review of the interventions to improve health and safety compliance.* Sudbury, UK: HSE Books.

Wright, M. et al. 2005. *An evidence based evaluation of how best to secure compliance with health and safety law.* Contract Research Report 334. Sudbury, UK: HSE Books.

Wright, P. 1977. Replies on reactor hazards "evasive". *The Times* 14 January p. 2.

Wu, T–C, Li, C–C, Chen, C–H and Shu C–M. 2008. Interaction effects of organizational and individual factors on safety leadership in college and university laboratories. *Journal of Loss Prevention in the Process Industries* 21(3), pp. 239–254.

Wustemann, L. 2011. *The Löfstedt review – a summary of recommendations and responses.* Retrieved from: www.healthandsafetyatwork.com/hsw/lofstedt-responses [Accessed 27 September 2016].

Yakhlef, A. 2007. Knowledge transfer as the transformation of context. *The Journal of High Technology Management Research* 18(1), pp. 43–57.

Young, D. 2010. *Common sense, common safety.* London: Cabinet Office.

Young, Lord. 2010. *Common sense, common safety [The Young Review].* London: Crown.

Zander, U. and Kogut, B. 1995. Knowledge and the speed of the transfer and imitation of organizational capabilities: an empirical test. *Organization Science* 6(1), pp. 76–92.

Zandvliet, R. 2011. Corporate social responsibility reporting in the European Union: Towards a more univocal framework. *The Columbia Journal of European Law Online* 18(38).

Zeleny, M., 2005. Knowledge-information autopoietic cycle: towards the wisdom systems. *International Journal of Management and Decision Making* 7(1), pp. 3–18.

Zeleny, M. 2006. From knowledge to wisdom: On being informed and knowledgeable, becoming wise and ethical. *International Journal of Information Technology & Decision Making* 5(4), pp.751–762.

Zohar, D. 2002a. The effects of leadership dimensions, safety climate, and assigned priorities on minor injuries in work groups. *Journal of Organizational Behavior* 23(1), pp. 75–92.

Zohar D. 2002b. Modifying supervisory practices to improve subunit safety: a leadership-based intervention model. *Journal of Applied Psychology* 87(1), pp. 156–163.

Zohar, D. and Luria, G. 2003. The use of supervisory practices as leverage to improve safety behaviour: a cross-level intervention model. *Journal of Safety Research* 34(5), pp. 567–577.

Zwetsloot, G. et al. 2008. Corporate social responsibility and psychosocial risk management. In: Leka S. and Cox, T. (eds.) *The European framework for psychosocial risk management: PRIMA-EF.* Nottingham, UK: I-WHO Publications. pp. 96–114.

# Index

Page numbers in *italics* denote a figure/table

Milton Keynes UK
Ingram Content Group UK Ltd.
UKHW030903141024
449569UK00026B/1315